PHYSICS of PHOTOREFRACTION in POLYMERS

Advances in Nonlinear Optics
A series edited by Anthony F. Garito, *University of Pennsylvania, USA*
François Kajzar, *DEIN, CEN de Saclay, France*
and Allan Boardman, *University of Salford, UK*

This book is part of a series. The publisher will accept continuation orders which may be cancelled at any time and which provide for automatic billing and shipping of each title in the series upon publication. Please write for details.

PHYSICS of PHOTOREFRACTION in POLYMERS

Dave West
The University of Manchester
Schuster Laboratory
Manchester, England

D.J.Binks
The University of Manchester
Department of Physics and Astronomy
Manchester, England

CRC PRESS

Boca Raton   London   New York   Washington, D.C.

Library of Congress Cataloging-in-Publication Data

West, Dave.
  Physics of photorefraction in polymers / by Dave West and David J. Binks.
    p.   cm. -- (Advances in nonlinear optics ; no. 6)
  Includes bibliographical references and index.
  ISBN 0-415-31073-3 (alk. paper)
    1. Refraction. 2. Polymers--Optical properties. 3. Photorefractive materials--Optical
  properties. I. Binks, David J. II. Title. III. Advances in nonlinear optics ; v. 6.

  QC426.P48 2004
  535′.324—dc22   2004054204

Visit the CRC Press Web site at www.crcpress.com

© 2005 by CRC Press

No claim to original U.S. Government works
International Standard Book Number 0-415-31073-3
Library of Congress Card Number 2004054204
Printed in the United States of America 1 2 3 4 5 6 7 8 9 0
Printed on acid-free paper

Contents

Introduction to the Series

Advances in Nonlinear Optics is a series of original monographs and collections of review papers written by leading specialists in quantum electronics. It covers recent developments in different aspects of the subject including fundamentals of nonlinear optics; nonlinear optical materials, both organic and inorganic; nonlinear optical phenomena such as phase conjugation, harmonic generation, optical bistability, fast and ultrafast processes, waveguided nonlinear optics, nonlinear magneto-optics, and waveguiding integrated devices.

The series will complement the international journal *Nonlinear Optics: Principles, Materials, Phenomena, and Devices* and is foreseen as material for teaching graduate and undergraduate students, for people working in the field of nonlinear optics, for device engineers, for people interested in a special area of nonlinear optics, and for newcomers.

Preface

Photorefractivity, the modulation of refractive index by the redistribution of photogenerated charge, was first reported in polymers little more than ten years ago. Today, photorefractive polymers are an exciting alternative to photorefractive crystals such as lithium niobate, with two notable economic advantages over their crystalline forebears: (i) straightforward material preparation with little waste and (ii) direct control during preparation of the features that will determine their photorefractive performance. Perhaps photorefractive polymers will succeed in making a commercial impact that has been difficult to achieve even with the many capabilities of photorefractive crystals.

The process of photorefraction in the most diffractive polymeric composites is analogous to the process in lithium niobate in that an optical pattern is reproduced as a space charge field in the medium. In crystals this field produces a phase hologram through the Pockels electro-optic effect. It was soon realized that here the polymers are different because, as is explained in this text, the statistical reorientation of anisotropic molecular polarizability within their amorphous structure is responsible for the production of a phase hologram. Photorefractivity in polymers also differs in numerous other ways from that in crystals. For example, the involvement of molecular reorientation within the amorphous medium leads to a wide dispersion in the holographic dynamics of these materials but there is little evidence that multilevel photorefractive charge trap models are needed.

There have been many empirical studies of different aspects of photorefraction in polymers published in the scientific literature. Much of this work has been based on assumptions of comparability between the effect in polymers and that of photorefractive crystals. This has been most notable in the study of holographic response times, which have tended to be analyzed empirically using notions of dispersive charge transport and multiple exponential dynamics with characteristic times. This work has led to conclusions such as that shallow traps must be present if a sublinear dependence is seen between the optical intensity and the holographic response rate.

At the University of Manchester and with Engineering and Physical Sciences Research Council (EPSRC) support, in the UK there has been over a decade of experimental and theoretical research studies in this field aimed at a more straightforward analysis. Accordingly, within this book we go back to the simple, single-trap level model of photorefraction as applied to polymers. This model of the growth of a space charge field is sufficient to explain all the known behaviors of this class of materials if the reorientation process is allowed to be dispersive, rather than insisting on a characteristic time scale. All unnecessary details have been removed in a bid to minimize the need in the model for material parameters that are not known. In this way we obtain a straightforward photorefractive polymer behavior that is predictable, for example, with linear dependence between recording optical intensity and rate of response of the space charge field. In this way, we hope that the need for empirical studies of this interesting class of materials will have passed.

The patterning of an electric field within a dielectric polymer has important potential applications in areas such as optical archiving and storage, optical information processing, and optical machine vision. To date, the physical models used to describe the process of photorefraction in polymers have been incomplete. Here we describe how a straightforward model can be devised to describe the static and dynamic properties of photorefractive polymers. We summarize the simple single-trap model of photorefraction as applied to polymers with high diffraction efficiency. There are chapters on the charge photogeneration process, the dispersion in charge transportation and the electro-optical response that creates a phase hologram from the pattern in the electric field. Finally, a chapter presents a simple model of the dispersion in the reorientational electro-optic dynamics of these materials, in which the number of free parameters to be determined from experiment has been reduced as far as is possible. The key feature of this model is that all aspects of the photorefractive response of highly diffractive photorefractive polymer composites can be considered in a self-consistent way and it is no longer necessary for empirical studies of correlations between parameters to be used.

About the Authors

Dr. West is lecturer in Applied Physics at the University of Manchester in the UK. His doctoral thesis at the University of Manchester described non-linear optical processes in polymeric Langmuir-Blodgett films. His early work on photorefractive polymer composites was supported by the Royal Commission for the Exhibition of 1851, and a series of grants from the Engineering and Physical Sciences Research Council (EPSRC), Defense Evaluation and Research Agency, UK (DERA), and industry have allowed the work to continue for almost 15 years. This work has attracted invitations to conferences and meetings in the United States, Europe, and Asia.

Dr. Binks is also a lecturer in the Department of Physics and Astronomy at the University of Manchester. Over a period of five years he has studied the physics and applications of photorefractive polymers, particularly the contribution of dipolar reorientation to index contrast dynamics. Prior to this, he studied for both his B.Sc. and Ph.D. degrees in Manchester, receiving them in 1992 and 1998, respectively.

List of Symbols

α	dispersion parameter; angle of resultant field to poling field; optical absorption coefficient
b	Onsager field parameter
β	angle of probe field to poling field
D	displacement vector; exciton diffusion coefficient
Δ	grating period
E	electric field, use subscript 0 to indicate spatially invariant component of this
E_D	diffusion field parameter
E_M	mobility field parameter
E_q	saturation field parameter
e	electronic charge
ε	dielectric constant
ϕ	angle of space charge field vector to poling field; quantum efficiency of photogeneration
Φ	phase angle of space charge field w.r.t. optical grating
g	rate of photogeneration of mobile hole density
γ_R	recombination constant
J	current density
K	grating vector
kT	product of Boltzmann constant and absolute temperature
l	exciton diffusion length
m_e	effective mass
μ	mobility
μ_{ij}	dipolar transition moment between i and j states
μ_{ii}	ith state diplar moment
N	number density of molecules
N_A	Avogadro constant; number density of acceptor molecules
N_A^-	number density of ionised acceptor molecules
N_{A0}^-	spatially invariant component of the number of density of ionised acceptor molecules
n	director vector
n	index of refraction
n_o	ordinary index of refraction
n_e	extraordinary index of refraction

n_p	prism index
n_{eff}	effective refractive index in a waveguide
n_2	nonlinear refractive index ($n = n_0 + n_2 I$)
v	wave frequency
P_n	Legendre polynomials
p	hole density; here the exponent of field dependence of photogeneration coefficient
p_0	spatially invariant component of hole density
$\mathbf{p}, \mathbf{p}_j$	density of molecular polarization
$\mathbf{P}, \mathbf{P}_j$	density of molecular polarization
ψ	wave phase ($= \omega t - \mathbf{kr}$ for propagation)
Q	transfer matrix
R	total reflectivity
r, r_{ij}	linear electro-optic, coefficient
R_{ij}	reflection matrix elements
r_{ij}	reflectivity on the interface between ith and jth layer
ρ	modulus of reflectivity ($r = \rho e^{i\phi}$); distance between hop sites
S	order parameter
s	coefficient of photoionization
s, s_{ijk}	quandratic electrooptic coefficient
σ	($= k_{zj}/k_{zi}$) ratio of k zth component between two layers (cross section)
T	temperature
T	total transmissivity
TE	transverse electric
TM	transverse magnetic
t	waveguide thickness
t_j	ith layer thickness
t	complex transmission; time
T_{ij}	transmission matrix elements
t_{ij}	transmission from ith to jth layer
t_{tr}	transit time
Θ	orientation angle
θ	propagation angle
$\mathscr{F}_{ij}$	transmission matrix elements
V	volume
ω	circular wave frequency ($= 2\pi v$)
Ψ	angle of probe wave electric vector to poling field
ξ_{1}	phase shift in layer i
ζ	azimuthal angle
<>	average
| |	modulus

chapter one

Why photorefractive polymer composites?

Light passes through many optical materials with no significant change in the properties of the wave other than an apparent reduction in the velocity (due to the refractive index). Often, in most uniform transparent materials, two light waves are able to pass through each other in the same volume of material without any interaction occurring between them. That this is so, leads naturally to the widespread success of the principle of linear superposition of the electric fields within transparent media, summarized as follows: under typical illumination conditions the electric fields of light waves are sufficient only to create minor polarizations of the material, so where two or more optical fields are present the effects of these tend to be independent of each other. There are, of course, materials in which this is not true such as where two or more optical fields can be made to interact by the effect of the material. One well-known example of this is the hologram.

In holography two coherent light waves are overlapped to create an interference pattern as in Figure 1.1.[1] A hologram is created when a finely grained, thick photographic emulsion is placed in the fringe pattern of light and darkness. After exposure the emulsion can be taken away and developed to create a pattern of transparency and opacity. The pattern created has the special property that if one of the light waves responsible for it is recreated then some of the light from that wave is diffracted to recreate the detail of the wavefronts of the second wave. This is why the familiar images with depth are observed from holograms. Coupling of energy from one wave to the other is due to the pattern that has been created in the hologram by the sequence of optical pattern exposure and chemical development.

With the development of lasers, there has been successful development of materials in which optical polarization itself is nonlinear so that no chemical development step is needed.[2] Typically a nonlinear material can lead, where the optical intensity is high, to coupling of one wave into another. A simple example is where a dye solution is used that is resonant with the wavelength of light used.[3] In this case a high intensity of light will begin to

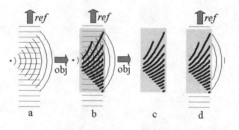

Figure 1.1 The principle of holography. (a) A reference wave is overlapped with a coherent object wave. (b) Holographic storage medium is placed in the region where interference between the two waves produces a fringe pattern. (c) The hologram so created exists even without the input light. (d) When the reference wave is again passed through, the hologram light is diffracted to recreate the object wave. (Diagram courtesy of Optitek.)

saturate the dye so that polarizability drops away from linearity with incident field.

Photorefractive materials have properties that distinguish them from those above. Like a holographic emulsion, they respond well to patterned illumination. But photorefractive materials do not require a separate chemical development step before a material response is obtained. Additionally a photorefractive material responds primarily to the way that light changes in intensity through the material, rather than the average intensity. If exposed to an optical pattern for a long time, then a steady state is approached in which a photorefractive material has maximized pattern contrast induced by the light. There is no overexposure process that destroys the pattern as is often found in photoreactive materials. The photorefractive material responds progressively to the fluence of light entering it. The pattern in the material is reinforced and develops in an asymptotic way toward its final steady state.

This property is very important, for example, where a material is to be used as a coherence gate in a full field optical imaging system.[4] Light scattered from an object is combined with coherent light from a mirror and a photorefractive material is placed in the fringe pattern that is formed, as in Figure 1.2. The hologram formed can be used to recreate only the image at a specific depth of the object, in a manner similar to optical coherence tomography but without the need to scan a point of light across the object. The light scattered from the other depths of the object can be excluded because the background scattered light that may be present will act only to reduce the holographic contrast slightly within the photorefractive material. It is fortunate that the pattern contrast in a photorefractive material is only weakly influenced by uniform background illumination (which can even accelerate the dynamic response of the material).

Patterns are stored in the material in the form of a nonuniform distribution of mobile charge. A simple change to a different illumination pattern

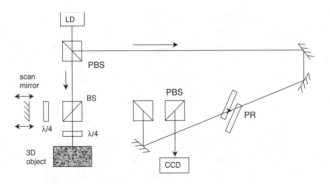

Figure 1.2 Principle of a depth-resolved imaging system to record and read whole-field holograms within a photorefractive polymer composite. Scatter from a 3D object is combined interferometrically with a reference wave from a scannable mirror that defines the depth to be resolved. An interference fringe pattern is created in a photorefractive device with a spatial amplitude pattern containing the image. The hologram created is used to diffract a readout wave incident on the back of the photorefractive into a CCD detector. LD, luminescent diode; BS, nonpolarizing beam-splitter; PBS, polarizing beam splitter; $\lambda/4$, quarter wave plate; PR, photorefractive composite device. In the Linnik variation identical microscope objectives are used to focus light onto the object and the reference mirror, in the sample and reference arms of the interferometer, respectively.

produces a new pattern. Uniform illumination by a single wave will erase the pattern. The material forms a pattern that can be modified an arbitrary number of times. In this way, these materials can be used again and again. This rewritable property has motivated studies of photrefractive materials for archival use in holographic optical storage of information with enormous information density (see Figure 1.3).[5,6]

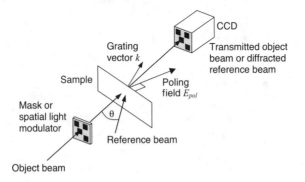

Figure 1.3 Schematic of a holographic storage optical arrangement showing the incoming object and reference beams and the corresponding transmitted and diffracted beams. During the recording process, an object beam is combined with a reference beam in the sample. During readout of the hologram, an object beam is not used, and the reference beam is diffracted into the detector.

The photorefractive effect was first identified in the electro-optic crystal lithium niobate where its consequence was first reported as a reversible form of optical damage to the transmission of an incident beam through the crystal. Holography using photorefractives has been proposed as a useful process with many potential applications and there have now been two decades of diverse optical research that has explored these.[7] One of the reasons photorefractive crystals have not realized commercial success is that the ferroelectric crystals that exhibit the effect are costly to grow, cut, and polish. The photorefractive effect is based on the impurity content in the crystals and this is difficult to control through a crystal boule as it is grown. For this reason, reproducibility of results can be difficult to attain when comparing different crystals in different situations.

For these reasons, it has been attractive to develop polymer composites that show the photorefractive effect. A polymer can be injection molded, spin coated into optical waveguide films, embossed into channels, and end facets can be produced with a quality dicing saw (as in Figure 1.4). There is no need for the costly crystal growth process or for polishing processes when a polymer can be used. Perhaps a polymeric photorefractive material could make commercial sense? An effort to produce polymeric photore-fractive materials led to the first reported success[8] at IBM Almaden Research Center.

After polymer composites exhibiting the photorefractive effect were developed, there were some unanticipated benefits. The optical pattern attained in a polymer through the reorientational effect is far higher in contrast than has been attained in crystals such as lithium niobate.[9] Optical

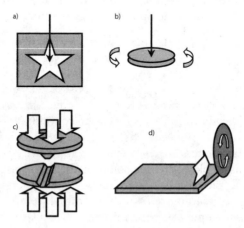

Figure 1.4 Polymers can be processed into many shapes and forms by processes including (a) injection molding of arbitrary shapes; (b) spin coating of thin films on a rotating substrate; (c) hot embossing of end facets, channels, or grooves; (d) end facet production using a high-quality dicing saw.

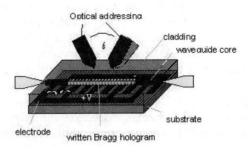

Figure 1.5 Schematic showing the basic principle of a reconfigurable Bragg grating in a photorefractive polymer waveguide. Two incident waves interfere to produce a fringe pattern to produce a Bragg grating. Changing the angle between the two incident waves alters the period of the grating induced. In this example, an interdigitated electrode pattern is used to produce electric poling field along the direction of the optical waveguide channel from left to right.

waves can be coupled into each other asymmetrically rather more strongly in a polymer than is the case in a crystal. It is much easier to develop polymer composites than it is to compare crystals of different dopant densities. The photosensitivity of the composites is easy to control by a poling electric field with a response of the sensitivity to field control that can be much greater than linear. This may lead to applications of photorefractive polymers in place of photosensitive waveguide or fiber materials where continuing reconfigurability is an advantage (Figure 1.5).

Physical model of photorefraction in polymeric composite media

Progress on photorefractive polymer composites was aided by the preexistence of the theory of photorefraction in crystals, but the areas where composites differ critically from their crystalline counterparts needed to be clarified.[10] The study of polymer composites was hindered initially by a relatively poor understanding in polymer composites of the process of photorefraction. This book addresses some of the novel aspects of polymeric media as photorefractive materials. The field dependence of charge photogeneration has consequences for both the dynamic and steady-state responses of the composites, for example. Perhaps, in the effort to understand, the most challenging issue has been the tendency of polymeric dynamics to be highly dispersive in their nature. Many dynamic responses in crystals are characterized by a temporal lifetime over which the first 66% or so of the approach to a steady-state response is obtained. But processes such as the movement of charges through the polymer and the reorientation of electric dipoles within the polymer do not exhibit characteristic timescales in this way.[11] The polymer is amorphous and each individual charge or dipole has a unique local environment. There is a distribution of dwell times for different locations and orientations. The difficulty has been to distinguish which of these dispersive processes are critical in improving the dynamic photorefractive response. All lead to nonexponential dynamics making multi-exponential fitting to obtain characteristic timescales unreliable. For this reason it is difficult to distinguish their individual consequences. The characteristic distributions of apparent response times may yet prove to be an advantage in some situations, for example where dynamic filtering of the low-frequency or high-frequency components of an evolving optical image are wanted.

Rather than attempt to describe the diverse and sometimes contradictory literature describing the discovery of photorefractive polymers and their behavior, this book presents a simple theoretical framework to understand their properties. Some insight is offered into the relative significance of the photogeneration process, the charge transport, and the dipole reorientation processes that are important in the most popular forms of photorefractive polymer composite. Where possible, the discussion is analytical, while the number of free parameters in the theory has been kept to a minimum. The following chapters therefore describe the simplest mathematics available to describe the behavior of photorefractive polymer composites. The analysis is both consistent with observations and as close as possible to the more mature analytical models developed for the photorefractive effect in crystal-line media such as lithium niobate, barium titanate, and the others.

References

1. Syms, R.R.A., *Practical Volume Holography*, The Oxford Engineering Science Series 24, Clarendon, Oxford, 1990.
2. Butcher, P.N., *The Elements of Nonlinear Optics*, Cambridge Studies in Modern Optics 9, Cambridge University Press, Cambridge, 1991.
3. Fisher, R.A., Ed., *Optical Phase Conjugation*, Quantum Electronics: Principles and Applications, Academic Press, New York; 1983.
4. Ansari, Z., Gu, Y., *et al.*, High Frame-Rate, 3-D Photorefractive Holography Through Turbid Media with Arbitrary Sources, and Photorefractive Structured Illumination, *IEEE J. Select Topics Quant. Electron.*, 7(6): 878–886, 2001.
5. Coufal, H.J., Psaltis, D., and Sincerbox, G.T., *Holographic Data Storage*, Springer Series in Optical Sciences, 76, Springer, Berlin, 2000.
6. Rahn, M.D., West, D.P., Khand, K., Shakos, J.D., and Shelby, R.M., Digital Holographic Data Storage in a High-Performance Photorefractive Polymer Composite, *Appl. Opt.*, 40(20): 3395–3401, 2001.
7. Solymar, L., Webb, D., and Grunnet-Jepson A., *The Physics and Applications of Photorefractive Materials*, Oxford University Press, Oxford, 1996.
8. Ducharme, S., Scott, J.C., Twieg, R.J., and Moerner, W.E., Observation of the Photorefractive Effect in a Polymer, *Phys. Rev. Lett.*, 66(14): 1846–1849, 1991.

9. Meerholz, K., Volodin, B.L., Sandalphon, Kippelen, B., and Peyghambarian, N., A Photorefractive Polymer with High Optical Gain and Diffraction Efficiency Near 100%, *Nature,* 371(6497): 497–500, 1994.

10. Moerner, W.E. and Silence, S.M., Polymeric Photorefractive Materials, *Chem. Rev.,* 94(1): 127–155, 1994.

11. Scher, H., Shlesinger, M.F., and Bendler, J.T., Time-scale Invariance in Transport and Relaxation, *Phys. Today,* 44(1): 26–34, 1991.

chapter two

Photorefraction in amorphous organic materials

This chapter presents a brief overview of the sequence of phenomena that are involved in efficient photorefractive response in polymer composites and glasses patterned by the creation of space charge fields. It particularly concerns those materials with low glass transition temperatures, at or near to ambient. The sequence of processes occurring in a photorefractive material can be summarized as:

- Light is absorbed and an ion-carrier pair is created.
- The ion is relatively immobile but the carrier can drift away under the influence of a field, toward a darker region.
- Mobile carriers are attracted to recombine with ions and the carrier population reaches a steady level quickly.
- Ions are left behind in the brighter regions while excess carriers collect in darker regions.
- An electric field builds up between bright and dark regions, which acts to oppose further accumulation.
- This electric field modifies the refractive index, leading to diffraction.

Typically, in amorphous organic materials, carriers are positively charged holes because the hole mobility is usually much greater than the electron mobility. There are only a few amorphous organic materials in which electron transport is dominant and as long as either one or the other form of transport is dominant the properties of the photorefraction process remain essentially similar. A standard model of photorefraction[1] has become accepted as the basis of our understanding of the process. This is described here for the simplest case;[2] a single kind of trap level in the majority case of hole transporting amorphous organic material.

The most publicized photorefractive polymer to date[3] consists of four different organic materials mixed together to form a composite. The polymer matrix is poly(N-vinylcarbazole), PVK, a well-known hole transport matrix used early on in the development of xerography (photocopying) with polymers.[4] PVK becomes a photoconductor in visible light when a small amount of 2,4,7-trinitro-9-fluorenone (TNF) is added. The photoexcitation of PVK-TNF leads to electron transfer from a side group of the PVK matrix, onto a stationary TNF molecule that becomes a (negatively charged) anion. A mobile electron vacancy (a hole) is left behind on the matrix. Either diffusion or, more importantly, drift in an applied field will move the hole through the matrix and as it moves the positive charge associated with it is attracted to other TNF anions fixed at various locations in the matrix, as in Figure 2.1. If the hole is captured by an anion then the two charges combine to produce a neutral TNF molecule as the hole is annihilated. The other two constituents of this most publicized photorefractive composite are an electro-optic dye and an additive that makes the material more plastic.

In this composite material, we can predict as a starting point that the rate of photogeneration of mobile holes should be proportional to the local intensity of light and the number density of neutral molecules of TNF, which acts as the photosensitizer. Charges will be generated nonuniformly if the intensity of light varies for different locations within the material. The charges are mobile and they will tend to spread throughout the material due to both diffusion and drift in a field.

The photogeneration of holes depends on the local intensity, whereas the annihilation of holes depends (in the single trap model) on the local number of anions. If the only source of anions was the photogeneration process, then the anion distribution would be patterned in just the same way as the intensity was patterned. When the illumination stops, all the holes remaining would recombine with the anions to erase the pattern, as in

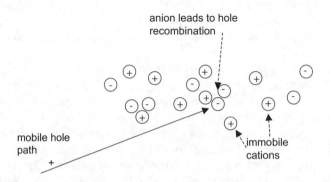

Figure 2.1 The movement of charge in an amorphous photorefractive hole transport matrix. Random distributions of immobile anions and cations coexist in the medium, trapped such that the recombination time is long. The smaller population density of mobile holes can be trapped by the anions anywhere within the medium.

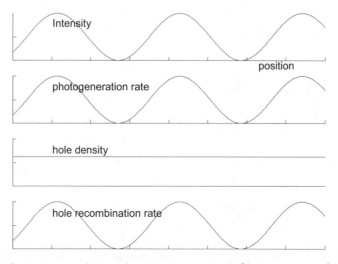

Figure 2.2 Generation and recombination patterns of photogenerated charges. The intensity pattern leads to an anion-hole pair photogeneration rate varying spatially. The mobile hole population spreads out due to drift and diffusion, so the recombination of holes with photogenerated anions is patterned in the same way as the generation of anions.

Figure 2.2. According to the single trap model, there would be no lasting effects in the material because we cannot distinguish between different holes.

So an important prerequisite for a useful photorefractive effect can be identified. The trap level must be filled partially before the photorefraction occurs. In this example, where the traps are the anions of the sensitizer, there must be a non-zero uniform population of anions for the optical patterning to take effect. When this is the case, then hole annihilation with anions will probably not leave the charge distribution unchanged because many carriers will annihilate at anions that were not created by the optical pattern, as in Figure 2.3.

The reduced contrast in the patterning of the number density of the anion of the sensitisor leads to a contrast similarly reduced in the rate of recombination between the hole and anion populations. The regions of high-intensity illumination become net sources of anion density whereas the darker regions are net sinks of anion density. The mobile hole population remains steady and small and the larger anion population becomes increasingly patterned as the illumination continues. The continuing presence of non-uniform illumination will make the contrast in the anion distribution grow and grow. The more anions that existed prior to illumination, the higher the population of traps and the greater the contrast in the anion distribution which can build up through capture of mobile holes in those traps. If there are enough traps (i.e., away from saturation) then the anion distribution will· eventually attain the same contrast as the optical pattern. When that has happened, there will be equilibrium between the processes of generation

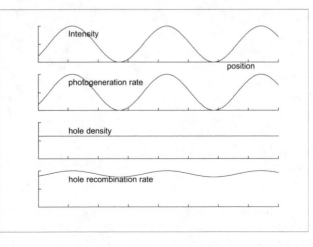

Figure 2.3 The effect of a significant uniform initial distribution of filled traps (in this case, sensitizer anions). The photogenerated anion population now patterns the recombination of holes with anions only slightly. In this case there can be a patterned inequality between photogeneration and recombination.

and loss of anions through the material. At this point the optical pattern has been recreated as an anion distribution in the material. The pattern in the anions will be displaced slightly from the optical pattern that created it, if a steady electric field was present during the photorefraction process. In the case of a sinusoidal intensity modulation, a spatial phase shift can be identified by which the anion pattern is displaced relative to the optical pattern. Such phase shifts are important in the coupling of light from one wave to another by a photorefractive hologram.

Electric charges act as sources of electric fields and the variation of anion density will lead to electric fields, which are maximized where the anion density is changing most rapidly. In the case of a sinusoidally varying optical intensity pattern the electric field due to the anion density pattern will have a $\pi/2$ phase displacement from the anion density pattern which created it. This is the reason why a photorefractive material exhibits a nonlocal response. The electric field is located away from the anion pattern that causes it, as in Figure 2.4.

The refractive index of the material must be modified by the patterned electric field due to the anion distribution to observe a photorefractive effect. A modified refractive index is obtained in poled polymer devices through the creation of asymmetry in the polarizability of an electro-optic dye distribution. A polar dye molecule has a positively charged region and a separate region of negative charge and a steady applied field can orient the relative positions of these charge regions, although the molecule itself remains fixed in space. Any additional changes in electric field will lead to polarization that depends on the direction of the new field. This can lead to a nonlinear response, for example a decrease in the field might lead to modest

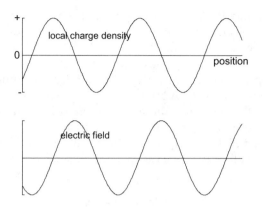

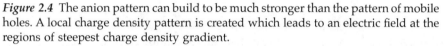

Figure 2.4 The anion pattern can build to be much stronger than the pattern of mobile holes. A local charge density pattern is created which leads to an electric field at the regions of steepest charge density gradient.

decreases in the polarization whereas an increase in the net field might lead to a proportionately greater rise in the polarization, as in Figure 2.5. An incoming optical wave will experience a polarizability that varies as the electric field reverses direction. The material has become non-centrosymmetric and it no longer possesses inversion symmetry. Within a Taylor power series expansion a new contributing term to the polarization appears which is dependent on the square of the electric field and this leads to the Pockels electro-optic shift in refractive index.

A more significant change in the refractive index is observed through the anisotropy of the dye molecules. Recall that the example composite material referred to earlier contained both an electro-optic dye and a plasticizer. The plastic environment of the composite has a low glass transition

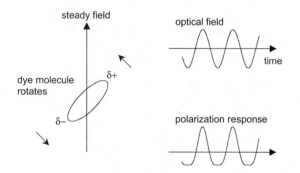

Figure 2.5 Asymmetric polarizability of a dye molecule: The molecules become aligned in a steady applied field. An optical field, E, incident onto the aligned molecular distribution then causes an asymmetric polarization which includes contributions dependent on E and on E^2. The Pockels electro-optic effect depends on the quadratic term.

temperature, T_g, relative to room temperature. The shear   viscosity of the individual dye molecules is reduced in the plastic environment to the point where dye molecules can rotate to respond to local changes in the static electric field. There will be more alignment of the dye molecules as the applied field becomes stronger relative to the randomizing effect of Brownian motion at room temperature. A high electric field leads to a strong tendency to alignment of the molecules because the energy of an oriented electric dipole in the electric field is large in comparison with the mean thermal energy, kT. The simplest description of the anisotropy of the refractive index of the molecules is that the index is given by

$$n = n_o + \Delta n \cos^2 \psi \qquad\qquad (2.1)$$

where n_0 is the smallest value and ψ is the angle between the dominant molecular axis and the probing optical electric field. The mean value of $\cos^2 \psi$ is, to a good approximation, proportional to the poling field squared.

Molecular reorientation therefore leads to a change in refractive index that is proportional to E^2, mimicking the Pockels electro-optic effect. Each dye molecule is itself highly anisotropic such that the polarizability is often much higher along the molecular axis than it is for electric fields across the molecular axis. In this case turning the molecule toward the electric field of a light wave increases the refractive index experienced by that wave, as shown in Figure 2.6.

The effect on refractive index of the local reorientation of the anisotropic polarizability profiles of the dye molecules due to a patterned electric field can be an order of magnitude greater than the effect of the Pockels electro-optic nonlinearity in a typical photorefractive composite with a low T_g.

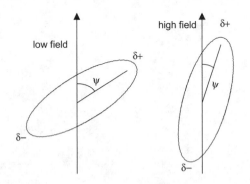

Figure 2.6 Alignment of an anisotropic molecule. Statistically, in a low field, the field is aligned poorly with the dominant molecular axis; polarizability and refractive index contributions are low. In a high field the molecular axis tends to be better aligned with the field leading to greater contributions to the polarizability and the refractive index.

For this reason, plasticized amorphous composites exhibit far stronger optical diffraction in photorefraction than do comparable materials without such freedom to reorient at the molecular level.

References

1. Solymar, L., Webb, D., and Grunnet-Jepson, A., *The Physics and Applications of Photorefractive Materials*, Oxford University Press, Oxford, 1996.
2. Kukhtarev, N.V., Markov, V.B., Odulov, S.G., Soskin, M.S., and Vinetskii, V.L., Holographic storage in electro-optic crystals. Beam coupling and light amplification, *Ferroelectrics* 22, 961–964, 1979.
3. Meerholz, K., Volodin, B.L., Sandalphon, Kippelen, B., and Peyghambarian, N., A photorefractive polymer with high-optical gain and diffraction efficiency near 100%, *Nature*, 371(6497): 497–500, 1994.
4. Weiser, G., *Phys. Stat. Sol. A.*, 18, 347–359, 1973.

chapter three

Model of the stored photorefractive hologram in amorphous organic media

The mathematical description of photorefraction in crystals has been established over two decades.[1,2] Although not every detail of this model has been verified yet, for the case of amorphous organic photorefractive materials, this standard model has naturally formed the basis of their development. The model is outlined here for the important case of an hole transport matrix with a single kind of trapping level. In the case of crystals it has been possible to refine this model as necessary to account for idiosyncrasies of the behavior of specific materials.[3] The nature of the dynamic of the electro-optic effect in organic materials with low T_g has inhibited the development of analogous multiple trapping models for the amorphous materials, for which the single-trap model of photorefraction remains the most important.

Charge photogeneration

Holes are generated with visible and infrared wavelengths of light through the photoionization of electronegative molecules that are referred to as sensitizers. The sensitizer molecules are assumed to be singly ionizable acceptors of electrons. The number density of acceptors not yet ionized is diminished by the presence of a small but significant residual acceptor ionization leading to an anion density population, N_A^-. The rate of photogeneration of mobile hole density is given by

$$g = sI(N_A - N_A^-) \tag{3.1}$$

where N_A is the total number density of acceptor molecules within the material (both ionized and neutral), I is the optical intensity and s is a constant of proportionality, the coefficient of photoionization. The rate at which the hole population is reduced due to annihilation with sensitizer anions will be dependent on the product of hole density and anion density, pN_A^-. The rate of change of the hole population, p, will obey the relation

$$\frac{\partial p}{\partial t} = sI(N_A - N_A^-) - \gamma_R pN_A^- \qquad (3.2)$$

where γ_R is a constant of proportionality, the recombination constant, that is dependent on the rate of Coulombic hole capture by the oppositely charged anions.[4] For the Langevin case of capture at singly charged sites $\gamma_R = e\mu/\varepsilon$ where the site charge, e, is within a medium of countercharge mobility μ and dielectric constant ε. Each photogeneration event produces an anion as well as a hole and the anion density, N_A^-, will change at the same rate as the density of mobile holes,

$$\frac{\partial N_A^-}{\partial t} = sI(N_A - N_A^-) - \gamma_R pN_A^- \; . \qquad (3.3)$$

Optical patterning

The photorefractive effect is a response to nonuniform illumination and the rate of photogeneration of charges will vary at different locations through the material during the process of recording a hologram. Equation 3.2 above is satisfied at each location through the material.

The anions are at fixed locations, whereas the hole population can move. Holes can diffuse away from regions of overcrowding where the intensity of light is high. When an applied electric field is present, holes will also drift in response to that field. A photocurrent can flow either as a real flux of charge carriers or as a displacement current as a displacement field is built up, as shown in Figure 3.1. Where the total photocurrent may be regarded as constant at all points within the material (neglecting current sources and sinks), the photocurrent density J will obey the continuity relation

$$J + \varepsilon_s \frac{\partial E}{\partial t} = J_0 = const. \neq f(\mathbf{r}) \qquad (3.4)$$

for all position vectors, $\mathbf{r}$, where $\varepsilon_s = \varepsilon_r \varepsilon_0$ is the static dielectric constant. The integral of this will be equal to the photocurrent measured in the circuit that applies the external field. The conduction current J_C due to hole motion may be written as

$$J_C = pe\mu E \qquad (3.5)$$

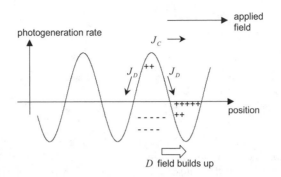

Figure 3.1 Three fates for a photogenerated hole. The applied field leads to drift current, J_C. The gradient of the hole density leads to a diffusion current, J_D, away from regions of high hole concentration. The motion of charges builds up a polarization corresponding to a displacement current, $\varepsilon_s \, \partial E/\partial t$.

where e is the magnitude of the electronic charge and hole motion has been described by a mobility, μ, which is assumed here to be the ratio of hole drift velocity/applied field strength. In amorphous organics the hole drift velocity is not independent of time because the transport process is dispersive. Hole drift velocity may be described by a mobility parameter, but in dispersive media the precise dynamics of hole transport will depend on the distance over which transport occurs. Using a single-valued mobility as is the case in crystals will be useful to assess the significance of mobility in the photorefractive process. It may lead, however, to difficulties when studying the dynamics of the holographic recording process.

The diffusion current J_D is proportional to the gradient of hole number density and to the thermal energy,

$$J_D = kT\mu(\nabla \cdot p) \tag{3.6}$$

as well as the mobility. In order to obtain acceptable efficiencies of charge photogeneration from incident light, it is typically necessary to apply a large electric field to an organic photorefractive device. Most organic materials to date are not cross-linked into a rigid structure because reorientational molecular freedom has been found to be beneficial (see chapter on electro-optics). Photovoltaic effects are not significant under these circumstances. Accordingly we can write that the external circuit photocurrent, J_0 from a photorefractive material

$$J_0 = pe\mu E + kT\mu(\nabla \cdot p) + \varepsilon_s \frac{\partial E}{\partial t} . \tag{3.7}$$

The divergence $(\nabla \cdot p) = \partial p/\partial z$ in the one-dimensional case of a sinusoidally modulated intensity pattern. This is valid for a geometry in which

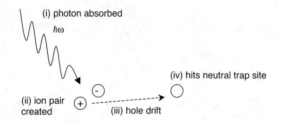

Figure 3.2 The Schildkraut model. Neutral traps capture holes to create the sensitizer anion population. The total hole trap density is then the total of anions and neutral donor sites. If the neutral trap sites are deep, these will form the passive donor sites (compensators) of the single-trap model of photorefractivity.

intensity is varying along one direction only (z) which corresponds to the case of interference planes due to two noncollinear coherent planar waves overlapping within the medium.

Electrical neutrality

The standard theory describes a process by which an optical pattern is recreated as a pattern of immobile ions. In the case of dominant hole transport of charge, there should be a significant and uniform distribution of anions prior to the application of the patterned intensity. If not, then recombination occurring only at photogeneration sites will not yield a persistent pattern in the anion distribution. The photorefractive medium is neutral overall and it follows from this that a significant and uniform distribution of immobile positive ions must also be present which coexists with the initial anion distribution such that $N_D^+ = N_A^-$. Either these donor cations play no role in the photorefractive effect, remaining charged and passive throughout the optical patterning,[5] or alternatively they play an early and short-lived role as neutral trap sites in the photorefractive process while the anion population is formed.[6,7] The standard model describes the behavior of the system during the subsequent time when the number density of donor cations is both uniform and invariant as the optical pattern is applied. Gauss' law of electrostatics gives the gradient of the electric field

$$\varepsilon_s (\nabla \cdot E) = -e(N_A^- - N_D^+ - p) \tag{3.8}$$

due to the patterning of the anion distribution.

Collective behavior of the photorefractive system

The simultaneous constraints on the photorefractive material of the anion density depends on the local optical intensity, that intensity variations leads to real and displacement currents, and that nonuniform charge density leads to space-charge field effects, all combine to determine the way that the system behaves. Mathematically, we have three coupled differential equations

$$\frac{\partial N_A^-}{\partial t} = sI(N_A - N_A^-) - \gamma_R p N_A^-$$

$$J_0 = pe\mu E + kT\mu(\nabla \cdot p) + \varepsilon_s \frac{\partial E}{\partial t} \qquad (3.9)$$

$$\varepsilon_s(\nabla \cdot E) = -e(N_A^- - N_D^+ - p)$$

in three dynamic variables of position and time; the electric field, E, the number density of holes, p and the number density of ionized sensitizer molecules, N_A^-.

The response of the material to the introduction of an optical pattern can be considered to be the sum of the response to the average intensity of the light and the effect of the pattern. The effect of the introduction of a uniform optical intensity, I_0, is described by the corresponding equations for the spatially invariant components

$$\frac{\partial N_{A0}^-}{\partial t} = sI_0(N_A - N_{A0}^-) - \gamma_R p_0 N_{A0}^-$$

$$J_0 = p_0 e\mu E_0 + \varepsilon_s \frac{\partial E_0}{\partial t} \qquad (3.10)$$

$$0 = -e(N_{A0}^- - N_D^+ - p_0)$$

where a subscript 0 has been added to indicate a spatially invariant component of a parameter. The donor population is considered to be a constant and therefore

$$\frac{\partial p_0}{\partial t} = \frac{\partial N_{A0}^-}{\partial t} \qquad (3.11)$$

because holes that are photogenerated leave behind corresponding anionic sites. Recombination reduces both the hole and anion populations similarly. Substituting (3.11) into the first equation of (3.10) we obtain

$$\frac{\partial p_0}{\partial t} = sI_0(N_A - N_D^+ - p_0) - \gamma_R p_0(N_D^+ + p_0) . \qquad (3.12)$$

The hole population is mobile and holes have a relatively short lifetime prior to recombination at anion sites. The hole population will not affect the electric field greatly because there are relatively few holes at any time. It is conventional to neglect the effect of holes on electric field using the assumption that $p \ll N_D^+$. If this assumption were not satisfied, then the photorefractive response would be weak unless additional or alternative trapping sites could be created. In the case that $p \ll N_D^+$, we find that an exponential growth of the hole population

$$p_0 = sI_0(N_A - N_D^+)\tau_h \left[1 - \exp\left(\frac{-t}{\tau_h}\right) \right] \tag{3.13}$$

is predicted due to the introduced uniform component of the optical illumination, where the hole recombination lifetime $\tau_h = (\gamma_R N_D^+)^{-1}$. In the long time limit of this dynamic, $t \gg \tau_h$, the hole population is equal to the product of the uniform component of the rate of hole photogeneration and the hole recombination lifetime.

The holes are the most mobile species in the amorphous photorefractive medium and their recombination with the embedded sites of opposite charge will be relatively rapid. The hole lifetime is accordingly one of the shortest of the timescales involved in the photorefractive process. The hole population is small and responds to intensity changes typically faster than the anion population, because the anion population is much larger. Note from the second equation (3.10) that in principle at least the photocurrent from the material is proportional to the hole population at that time, if the hole mobility is a constant.

The effect of the pattern in the intensity of light is to introduce patterning in both the hole and anion populations. A space charge pattern is created and an electric field pattern results from this. An analytical solution is possible in the linearized regime where the intensity pattern is not too high in contrast. Considering the one-dimensional case that the intensity, I, is the real component of the hypothetical complex parameter

$$\tilde{I} = I_0 + \Delta\tilde{I}\exp(-jKz) \tag{3.14}$$

where the grating vector $K = 2\pi/\Lambda$ and Λ is the period of the intensity pattern, we can look for solutions of the form

$$\tilde{E} = E_0 + \Delta\tilde{E}\exp(-jKz)$$

$$\tilde{p} = p_0 + \Delta\tilde{p}\exp(-jKz) \tag{3.15}$$

$$\tilde{N}_A^- = N_{A0}^- + \Delta\tilde{N}_A^-\exp(-jKz).$$

Under the assumption that $\Delta\tilde{I} \ll I_0$ the equations (3.9) are linearized by neglecting the second spatial harmonic terms in $\exp(-j2Kz)$ due to the product of modulation terms, to yield the approximate relation for the cross-product

$$\tilde{I}\tilde{N}_A^- \approx I_0 N_{A0}^- + (I_0\Delta\tilde{N}_A^- + \Delta\tilde{I}N_{A0}^-)\exp(-jKz) \tag{3.16}$$

with similar relations for the products $\tilde{p}\tilde{N}_A^-$ and $\tilde{p}\tilde{E}$. The three, coupled differential equations describing the anion density changes, the total current and the electric space-charge field can now be rewritten in the linearized regime for the terms in $\exp(-jKz)$ as

$$\frac{\partial}{\partial t}\tilde{N}_A^- \approx s\Delta\tilde{I}(N_A - N_D^+) - sI_0\Delta\tilde{N}_A^- - \gamma_R(p_0\Delta\tilde{N}_A^- + \Delta\tilde{p}N_{A0}^-)$$

$$0 \approx e\mu(p_0\Delta\tilde{E} + \Delta\tilde{p}E_0) + kT\mu\frac{\partial}{\partial z}\Delta\tilde{p} + \varepsilon_s\frac{\partial}{\partial t}\Delta\tilde{E} \tag{3.17}$$

$$\varepsilon_s\frac{\partial}{\partial z}\Delta\tilde{E} \approx e\Delta\tilde{N}_A^- .$$

These may be solved by noting that

$$\Delta\tilde{N}_A^- = -j\left(\frac{\varepsilon_s K}{e}\right)\Delta\tilde{E} \tag{3.18}$$

and thus $\Delta\tilde{N}_A^-$ can be eliminated from the top two equations. The approximations that $N_{A0}^- \approx N_D^+$ and that for times $t \gg \tau$ the average hole population density has reached a steady, saturated value

$$p_0 = \frac{sI_0(N_A - N_D^+)}{\gamma_R N_D^+} \tag{3.19}$$

are used to obtain a dynamic equation for the patterning of the electric field over time that has the form

$$A\frac{\partial}{\partial t}\Delta\tilde{E} + BA\tilde{E} = jC . \tag{3.20}$$

In order to describe the coefficients A, B, C of this equation, we introduce three field parameters by convention. The diffusion field parameter

$$E_D = \frac{kT}{e} K \tag{3.21}$$

is the electric field strength at which the potential drop across one radian of the intensity pattern is equal the mean thermal energy. The mobility field parameter

$$E_M = (\mu \tau_h K)^{-1} \tag{3.22}$$

is the electric field necessary for a hole to drift, on average, one radian within one recombination lifetime. The saturation field parameter

$$E_q = \frac{e N_D^+}{\varepsilon_s K} \left(1 - \frac{N_D^+}{N_A} \right) \tag{3.23}$$

is the maximum electric field strength which can be sustained within the material due to optical patterning of the charge distribution. This is the field modulation obtained when the anion population reaches maximum contrast. Using these three field parameters the coefficients are expressed according to the standard solution as

$$A = \frac{\varepsilon_s}{p_0 e \mu} \left(1 + \frac{E_D + jE_0}{E_M} \right)$$

$$B = 1 + \frac{E_D + jE_0}{E_q} \tag{3.24}$$

$$C = \left(\frac{\Delta \tilde{I}}{I_0} \right) (E_D + jE_0)$$

and with these parameters the growth of the pattern in the electric field follows the dynamic

$$\Delta \tilde{E} = j \frac{C}{B} \left[1 - \exp \left(\frac{-Bt}{A} \right) \right]. \tag{3.25}$$

The amplitude of the field patterning, $|\Delta\tilde{E}|$, rises nearly exponentially toward a saturation value in the long time limit given by

$$|\Delta\tilde{E}|_{t\to\infty} = \left|j\frac{C}{B}\right| = \left|j\left(\frac{\Delta\tilde{I}}{I_0}\right)\frac{E_q(E_D + jE_0)}{E_q + E_D + jE_0}\right| \tag{3.26}$$

and the reason the rise is nearly exponential is that the characteristic lifetime, τ_{rise}, of the buildup of this space-charge field pattern is complex and is given by

$$\tau_{rise} = \frac{A}{B} = \frac{\varepsilon_s}{p_0 e\mu}\frac{(E_M + E_D + jE_0)}{E_M}\frac{E_q}{(E_q + E_D + jE_0)}. \tag{3.27}$$

Note that both the mobility field parameter, E_M and the saturation field parameter, E_q, are proportional to the trap density, N_D^+ and materials with a very high-trap density will come close to an exponential rise in space charge field. The initial rate of rise of the pattern in the space charge field is constrained to be within the limit given by the modulus

$$\left|\frac{\partial}{\partial t}\Delta\tilde{E}\right| \leq \left|j\frac{C}{A}\right| = \frac{p_0 e\mu}{\varepsilon_s}\left|j\left(\frac{\Delta\tilde{I}}{I_0}\right)\frac{E_D + jE_0}{\left(1 + \dfrac{E_D + jE_0}{E_M}\right)}\right|. \tag{3.28}$$

Example:
In a typical experiment, an intensity pattern with a period of 3 μm is used and at room temperature the diffusion field $E_D = 10^4$ Vm^{-1}. This is significantly smaller than the poling field used in the range $E_0 = (5 - 10) \times 10^7$ Vm^{-1} or the saturation field of a typical material, $E_q \geq 5 \times 10^7$ Vm^{-1}. The approximation that $E_D << E_0, E_q$ is usually a good one and in this approximation the field pattern saturates at a limiting value of

$$|\Delta\tilde{E}|_{t\to\infty} \approx \left|\frac{-E_q E_0}{(E_q + jE_0)}\frac{\Delta\tilde{I}}{I_0}\right| \tag{3.29}$$

while the initial rate of rise of the space-charge field

$$\left|\frac{\partial}{\partial t}\Delta\tilde{E}\right| \leq \frac{p_0 e\mu}{\varepsilon_s}\left|\frac{-E_M E_0}{(E_M + jE_0)}\frac{\Delta\tilde{I}}{I_0}\right| \tag{3.30}$$

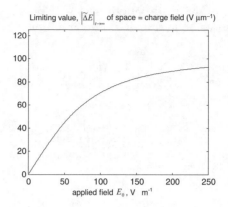

Figure 3.3 Space-charge field in the steady state as a function of applied field strength. A saturation field parameter $E_q = 100$ V μm⁻¹ was assumed and the diffusion field was neglected. Although applied fields up to 250 V μm⁻¹ are shown, it is difficult to hold more than 100 V μm⁻¹ across a typical material for long without breakdown.

and these two very similar relations are plotted for example material parameters in Figures 3.3 and 3.4.

Re(τ_{rise}) is a characteristic time for the approach to the steady-state electric field response of the photorefractive material to a patterned optical stimulus. The imaginary component of τ_{rise} represents an oscillatory term $\exp(jt/\tau_{rise})$ in the response dynamic for the space charge field. Oscillations are not normally observed in the transient for the diffraction from holograms when they are recorded in organic composite photorefractive media with a low T_g. This is due probably to the smoothing effect of the reorientational mechanism of electro-optic response, in which the refractive index depends strongly on the orientational distribution of the electro-optic dye chromophores. The

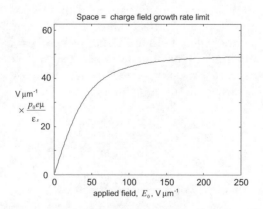

Figure 3.4 Applied field dependence of the rate limit to the growth of the space charge field. A mobility field $E_M = 50$ V μm⁻¹ was assumed and the diffusion field was neglected.

imaginary component of the rise time is responsible for the gain enhancement that has been observed in two beam coupling experiments with a
moving optical pattern on photorefractive polymers.[8]

Trap density

The term *trap* is applied in the context of photorefractive materials normally
to sites that capture mobile charge temporarily, allowing the carriers to be
released optically within the duration of a typical experiment. In the single-trap level model, the traps are the population of ionized sensitizer molecules, with number density $N_A^- \approx N_D^+$. The density of trap sites determines
the holographic resolution with which an optical pattern may be reproduced
as a pattern in the electric field within a photorefractive material, in a manner
crudely analogous to the grain size of a fast photographic film. The saturation
field in (3.23) is dependent on the ratio of trap density to spatial frequency
of the pattern, N_D^+ / K. Alternatively, the saturation field depends on the
product of trap density and spatial period of pattern, $N_D^+ \Lambda$. Small details of
the pattern will be lost if the trap density is insufficient to resolve them. The
analogy with grain size is less accurate when considering the rate of recording of a hologram within a photorefractive material. Both the mobility field
and the saturation field are proportional to the trap density and the rise time
of the holographic space charge field may be almost unchanged as the trap
density changes.

   The trap density has a major effect on the spatial phase shift of the electric
field pattern relative to the optical pattern that created it. The phase of the
electric field modulation is displaced from the intensity pattern by the phase
angle, Φ, of the complex factor jC/B and in the useful approximation that
we neglect the diffusion field parameter $E_D << E_0, E_q$ this reduces to

$$\Phi = \arctan\left(\frac{imag(jC / B)}{real(jC / B)} \right) = \arctan\left(\frac{-E_0}{E_q} \right). \tag{3.31}$$

   The phase shift between the intensity pattern and the electric field modulation that results from it is found to be dependent on the saturation field,
E_q and therefore the trap density (Figure 3.5).

   In some cases, there may not be permanently charged donor sites within
an amorphous photorefractive medium, if the charged state of the donor is
metastable, but not permanent.[9] In these cases the dominant trap sites will
be uncharged initially. If these uncharged trap sites are deep enough in
energy then optical de-trapping will not occur on experimental time scales
and the traps will become filled during the early period of a photorefractive
recording process. After the traps are filled, by acting as donors to trap mobile
holes, they then act precisely as the permanent donor cations, or compensators, of the single-trap model of photorefraction. This model has been studied

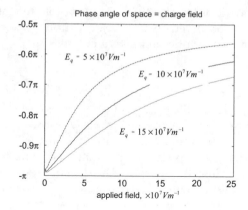

Figure 3.5 Trap density determines the grating phase. Phase angle, Φ, of space-charge field modulation relative to optical pattern, for different applied field and saturation field strengths. The saturation field is proportional to the ratio of trap density to spatial frequency of the pattern, N_D^+/K. A diffusion field ED = 5×10^4 Vm^{-1} has been assumed.

analytically. The main effect of this behavior then is the introduction of two distinct regimes of operation of a photorefractive medium. At first the neutral trap sites dominate the trapping process and the second regime follows when these neutral donor sites have become positively charged.

There is some evidence of instances of uncharged trap site effects. High correlation has been observed between the trap density in photorefractive studies and the intensity of the spectroscopic signature of the C_{60}-anion.[10] This was explained by the authors as a result of the creation of anion and cation populations during uniform pre-illumination. Those authors speculated that the chromophores formed the uncharged hole traps. A temperature dependence of the trap density in photorefractive studies was observed near T_g in a composite sensitized by TNF rather than C_{60}.[11] The softer the composite became, the fewer trap sites were present and this was explained by the notion that neutral traps are formed by the extreme tail of the Gaussian distribution of energy levels involved in the hole transport process. The inhomogeneous broadening of the energy levels is considered to be due at least in part to conformational disorder, which is expected to decline as the composite softens. Another possibility in the case of PVK is the formation of carbazole group dimer states that act as trap sites.[12]

Electro-optic response in low T_g materials

Both the Pockels electro-optic effect and the reorientational effect due to the anisotropic polarizability of the dye molecules lead to a quadratic dependence of the refractive index on the electric field,

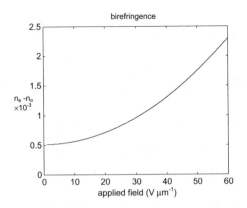

Figure 3.6 A typical example of the birefringence induced by an applied field on a low T_g amorphous photorefractive composite. These data are from ellipsometry on the composite 51.5 %wt. PVK with 47.5 %wt. EHDNPB and 1 %wt. TNF. The zero field birefringence is due to stress.

$$n = n\big|_{E=0} + \Delta n(E^2) \, , \tag{3.32}$$

as shown in Figure 3.6. The field will induce birefringence in the medium, changing refractive index differently for optical polarizations parallel and perpendicular to the applied field. An optical probe wave polarized with electric vector at an angle ψ to the applied poling field will see a refractive index, n_p, given by the index ellipsoid equation [2,13,14,15]

$$n_p^{-2} = n_e^{-2} \cos^2 \psi + n_o^{-2} \sin^2 \psi \tag{3.33}$$

as shown in Figure 3.7. Here n_e and n_o are the extraordinary and ordinary refractive indices, the two limiting cases of refractive index for light with electric vector parallel and perpendicular to the poling field, respectively.

The birefringence seen by the probe wave, correct to first order in $\Delta n/n = (n_e - n_o)/n_o$ is

$$n_p - n_o = (n_e - n_o) \cos^2 \psi \, , \tag{3.34}$$

which is the difference between the refractive index for the two most extreme cases of linear polarization of the probe wave. n_p is the index experienced when the probe wave is polarized with electric vector in the plane of incidence and diffraction, referred to as p-polarization, as in Figure 3.8. This corresponds to the maximization of the dot product of the poling field and optical field vectors without a change in the propagation direction, ψ, of the

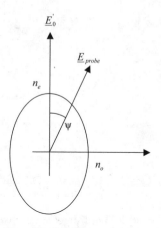

Figure 3.7 For a single steady applied field on a photorefractive composite with low T_g, the index ellipse is a radial plot of the refractive index as a function of angle, ψ, between the electric vectors of the static applied and oscillating probe wave fields.

probe wave. The refractive index, n_o, is experienced for a wave traveling in the same direction but with a perpendicular polarization referred to as s-polarization (with electric vector perpendicular to both the poling field and the plane of incidence and diffraction). For s-polarization the dot product of poling field and optical field vectors is zero.

In general both the extraordinary and ordinary refractive indices will depend quadratically on the local applied field, with the extraordinary index (parallel to the poling field) most strongly affected. For s-polarized light, $\psi = \pi/2$ and the behavior of Pockels and reorientational electro-optic response mechanisms differ. For the Pockels effect the change in index for s-polarization

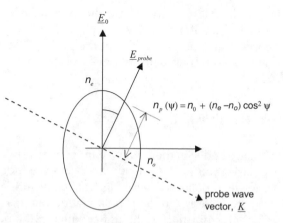

Figure 3.8 The optical probe wave experiences a refractive index given by the transverse dimension of the index ellipse.

$$\left.\Delta n_o\right|_{Pockels} = +\frac{1}{3}\Delta n_e \tag{3.35}$$

while the reorientational effect leads to an index change for s-polarization

$$\left.\Delta n_o\right|_{reorient.} = -\frac{1}{2}\Delta n_e \tag{3.36}$$

and in typical photorefractive composites with low T_g, it is the reorientational effect which dominates. The refractive index experienced by the p-polarized probe wave can be written in the case of the Pockels effect as

$$\left.n_p\right|_{Pockels} = n + (n_e - n_o)\left[\cos^2\psi + \frac{1}{2}\right] \tag{3.37}$$

where n is the isotropic refractive index of the medium in the absence of the field. This is valid for rigid materials with a high T_g where negligible reorientation of dye molecules occurs at room temperature. In the dominance of the reorientational effect the index becomes

$$\left.n_p\right|_{reorient.} = n + (n_e - n_o)\left[\cos^2\psi - \frac{1}{3}\right] . \tag{3.38}$$

Photorefractive space-charge field

The photorefractive response creates a space-charge field that is modulated in proportion to the optical pattern,

$$\text{Re}\left(\Delta\tilde{E}\exp(-jKz)\right) = E_{sc}\cos(Kz + \Phi) , \tag{3.39}$$

as shown in Figure 3.9.

In the absence of a poling field, the refractive index in a low T_g material is proportional to the square of this field,

$$\Delta n(E^2) \propto \text{Re}\left((\Delta\tilde{E})^2\exp(-j2Kz)\right) . \tag{3.40}$$

The hologram in these circumstances is formed with twice the spatial frequency of the optical pattern and Bragg diffraction from such a hologram will occur at anomalous angles. The hologram due to space-charge field alone is not an accurate reproduction of the optical pattern.

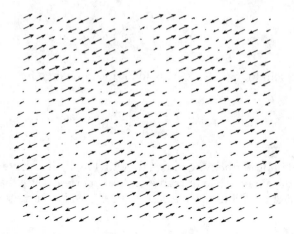

Figure 3.9 Photorefractive space-charge field for a grating vector tilted 60° clockwise from vertical.

The combination of the space-charge field with a poling field is important where hologram fidelity to the recording optical pattern is required. In a photorefractive hologram the poling field is changed locally by the space-charge electric field created through the photorefractive effect when under nonuniform illumination. The space-charge field will be directed along the steepest gradient in the optical intensity pattern. With a unidirectional holographic grating vector, this corresponds to the grating vector of the sinusoidal intensity pattern. In general the space-charge field will not be parallel to the steady applied poling field, E_0', as shown in Figure 3.10.

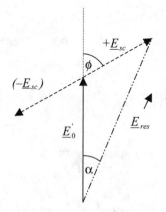

Figure 3.10 The arrangement of electric field vectors in a hologram. The poling field, E_0', combines with the space-charge field, $\pm E_{sc}$, to give a resultant field, E_{res}, which changes in both magnitude and direction, α, as the space-charge field varies through the hologram.

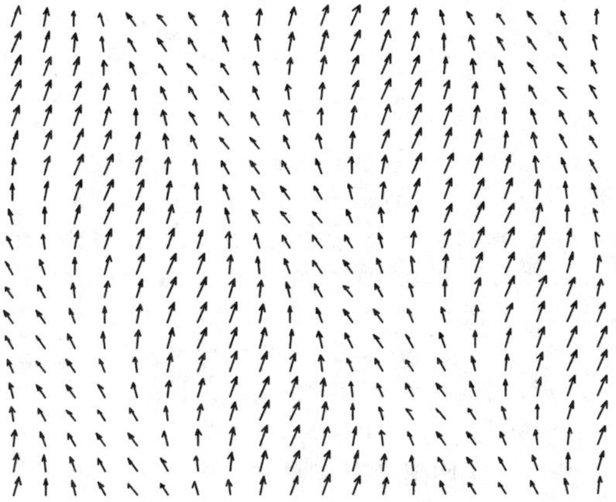

Figure 3.11 The pattern of resultant field corresponding to the space-charge field of Figure 3.9 with the addition of a strong poling field oriented vertically upward.

The holographic fidelity is far better when a strong poling field is applied. The total field, equal to the sum of the poling field and the space-charge field, changes locally in magnitude and orientation as the space-charge field varies through a photorefractive hologram, as in Figure 3.11.

A resultant electric field

$$E_{res} = E'_0 + E_{sc} \cos(Kz + \Phi) \tag{3.41}$$

is produced and the square of this field is given by cosine law as

$$E_{res}^2 = E_{res} \cdot E_{res} = (E'_0)^2 + E_{sc}^2 \cos^2(Kz + \Phi) + 2E' \cdot E_{sc} \cos(Kz + \Phi) \tag{3.42}$$

where the existence of the accurate field modulation with grating vector K depends on a non-zero projection of the grating vector along the poling field. The accurate field modulation term appears with amplitude $2E'_0 E_{sc} \cos \phi$, where ϕ is the angle between the grating vector and the poling field. The angle, α, between the resultant field direction and the poling field direction is given by

$$\cos \alpha = \frac{E'_0 + E_{sc} \cos \phi \cos(Kz + \Phi)}{E_{res}} \tag{3.43}$$

and this angle varies through the hologram.

In chapter 6, it is shown that the refractive index experienced by light with electric vector parallel to the poling field will vary according to both the magnitude and orientation of E_{res} such that the electro-optic index change is proportional to

$$\Delta n_e(z) \propto E_{res}^2 \left[\cos^2 \alpha + \frac{\Delta n_o}{n_e - n_o} \right] \tag{3.44}$$

where the latter ratio of ordinary index change to birefringence is a material constant and the index contrast for the probe wave depends on a field-squared which is modified geometrically due to the room temperature reorientation of chromophores. In reorientational materials with a low T_g, the change in ordinary index is, $\Delta n_o \approx (1/3)(n_e - n_o)$.

The probe wave used to observe diffraction from the hologram is not, in general, polarized with electric field parallel to the poling field, but at an angle, β, to it. In this case the electro-optic index change, $\Delta n_p(z)$, experienced by the probe wave will be proportional to

$$\Delta n_p(z) \propto E_{res}^2 \left[\cos^2(\alpha - \beta) + \frac{\Delta n_o}{n_e - n_o} \right] \tag{3.45}$$

for a probe wave field, a poling field and a space-charge field which all lie on the same plane (the case of p-polarization). Substituting for α and rearranging yields the proportionality

$$\Delta n_p(z) \propto \left(E_0' \cos \beta + E_{sc} \cos(\phi - \beta) \cos(Kz + \Phi) \right)^2 + \left(\frac{\Delta n_o}{n_e - n_o} \right) E_{res}^2 \tag{3.46}$$

describing how the holographic contrast in refractive index will depend on the orientations of the space-charge field (angle ϕ) and the probe wave field (angle β) relative to the poling field.

The long time asymptotic limit to the space-charge field is given by the projection, $E_0 = E_0' \cos \phi$, of the poling field along the grating vector at an angle ϕ from the poling field direction,

$$\left| \Delta \tilde{E} \right|_{t \to \infty} \approx \left| \frac{-E_q E_0' \cos \phi}{(E_q + j E_0' \cos \phi)} \frac{\Delta \tilde{I}}{I_0} \right|, \tag{3.47}$$

and if the trap density is sufficient to avoid trap saturation, $E_q \gg E_0$, then the approximate relation

$$\left| \Delta \tilde{E} \right|_{t \to \infty} \approx m E_0' \cos \phi \tag{3.48}$$

is useful where the factor $m = |\Delta \tilde{I}/I_0|$ is the scalar contrast in the optical pattern present.

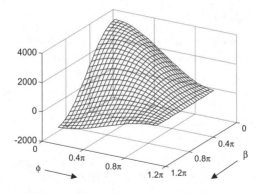

Figure 3.12 The first Fourier component of the modulated squared field (in $V^2 \mu m^{-2}$) as experienced by a probe wave field at angle β, through a space-charge field with angle ϕ, both relative to a poling field of magnitude 50 $V\mu m^{-1}$ (see (3.49)). The ratio $\Delta n_o/(n_e - n_o) = -0.3$ in this calculation, corresponding to a strongly anisotropic dye within a reorientational composite. This calculation has been based on the Kuhktarev model and reorientational enhancement of the electro-optic response, but it does not allow for modifications of the standard theory of photorefraction such as field-dependent efficiency of charge photogeneration.

Thus, according to the standard model of Kuhktarev and simple theory of reorientational enhancement of electro-optic response, the first Fourier spatial component of the holographic contrast (3.46) is governed by

$$\Delta n_p\big|_{1K} \propto 2E_0 E_{sc} \left[\cos(\phi - \beta)\cos\beta + \frac{\Delta n_o}{n_e - n_o}\cos\phi \right] \qquad (3.49)$$

and this function is plotted in Figure 3.12.

The expression (3.49) above for the contrast for p-polarization shows how the accurate (1K) hologram has an optimum contrast when ϕ and β are both zero and all fields are parallel. As the space-charge field is rotated away from the poling field the contrast falls, reaching zero for space-charge field perpendicular to the poling field, $\phi = \pi/2$. As the probe wave field is rotated away from the poling field, the holographic contrast also falls. Note that if the space-charge field is nearly parallel to the poling field ($\phi \approx 0$), then even a probe field perpendicular to the poling field can experience a true 1K hologram, which is shifted in phase by π, in this case for a reorientational (low T_g) material.

The corresponding expression for the index contrast for probe light with linear polarization perpendicular to the plane of the poling field and the ray paths (s-polarization) is more simple. At all locations s-polarized light will have $\alpha - \beta = \pi/2$ and the contrast in this case

$$\Delta n_o\big|_{1K} \propto 2E_0' E_{sc} \frac{\Delta n_o}{n_e - n_o}\cos\phi \; . \qquad (3.50)$$

Diffraction efficiency

According to Kogelnik's formula[16] for the diffraction efficiency from a hologram the diffraction efficiency for p-polarized light

$$\eta = \sin^2\left(\frac{\pi\Delta nL}{\lambda}\cos\Delta\beta\right) \tag{3.51}$$

depends on the product of the holographic contrast and the cosine of the
scatter angle between the probe and the diffracted waves, $\cos\Delta\beta$. (L is the
interaction length within the medium and λ is the wavelength of light used.)
The factor $\cos\Delta\beta$ is needed because the material polarization due to the
probe wave field is not oriented precisely transversely in the frame of the
diffracted wave. In the frame of reference of the scattered wave, only the
transverse component of the probe polarization can radiate into the diffracted wave, as shown in Figure 3.13.

In the derivation of the proportionality (3.49) above, the refractive index
has been considered to be a scalar quantity — the assumption has been
implicit that the polarization due to the optical probe is parallel to the electric
vector of that wave. The fact that the steady electric field has a component
along the direction of propagation of the probe wave suggests that a component of polarization should exist which is longitudinal in the frame of
reference of the probe beam (Figure 3.14).

This longitudinal polarization component would be expected to be weak.
However, if the light diffracted is deflected significantly, it could contribute
to the radiation into the diffracted wave. [17] This aspect is discussed in more
detail in chapter 6.

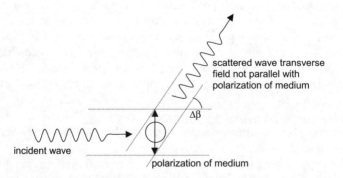

Figure 3.13 The material polarization induced by the probe optical field is not oriented correctly for efficient radiation into the scattered wave when p-polarized light
is used. Only the component of the polarization that is transverse to the scattered
wave will radiate into that wave.

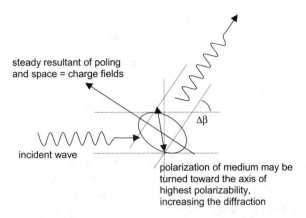

Figure 3.14 The anisotropy in polarizability due to the steady resultant field may lead in principle to longitudinal material polarization in the frame of the incident wave.

It is also necessary with organic composites to consider a field-dependent efficiency of carrier photogeneration when calculating the magnitude of the space-charge field. If the field dependence of the photogeneration efficiency is described by a power law relationship that $s \propto E^p$, then the space-charge field will saturate at

$$E_{sc}\Big|_{s \propto E^p} \leq \frac{E_0' \cos \phi}{1 + p \cos^2 \phi} \qquad (3.52)$$

if the trap density is sufficient and diffusion is negligible. The value of the exponent p varies according to the electric poling field strength used (see Figure 3.15). This aspect of photorefraction in organic media is discussed in chapter 4.

References

1. Kukhtarev, N.V., Markov, V.B., Odulov, S.G., Soskin, M.S., and Vinetskii, V.L., Holographic storage in electro-optic crystals. Beam coupling and light amplification, *Ferroelectrics*, 22: 961–964, 1979.
2. Yeh, P., *Introduction to Photorefractive Nonlinear Optics*, Wiley, New York, 1993.
3. Solymar, L., Webb, D., and Grunnet-Jepson, A., *The Physics and Applications of Photorefractive Materials*, Oxford University Press, Oxford, 1996.
4. Khand, K., Binks, D.J., West, D.P., and Rahn, M.D., Photorefractive trapping and the correlation between recording and erasure dynamics in a polymer composite, *J Mod. Opt.*, 48(1): 93–101, 2001.
5. Binks, D.J., West, D.P., Norager, S., and O'Brien, P., Field-independent grating formation rate in a photorefractive polymer composite sensitized by CdSe quantum dots, *J. Chem. Phys.*, 117(15): 7335–7341, 2002.

6. Schildkraut, J.S. and Buettner, A.V., Theory and simulation of the formation and erasure of space-charge gratings in photoconductive polymers, *J. Appl. Phys.*, 72: 1888–1893, 1992.
7. Schildkraut, J.S. and Cui, Y., Zero-order and first-order theory of the formation of space-charge gratings in photoconductive polymers, *J. Appl. Phys.*, 72: 5055–5060, 1992.
8. Grunnet-Jepson, A.C., Thompson, L., and Moerner, W.E., Gain enhancement by moving gratings in a photorefractive polymer, *Opt. Comm.*, 145(1–6): 145–149, 1998.
9. Silence, S.M., Bjorklund, G.C., and Moerner, W.E., Optical trap activation in a photorefractive polymer, *Opt. Lett.*, 19(22): 1822–1824, 1994.
10. Grunnet-Jepson, A., Wright, D., Smith, B., Bratcher, M.S., DeClue M.S., Siegel J.S., and Moerner, W.E., Spectroscopic determination of trap density in C_{60}-sensitized photorefractive polymers, *Chem. Phys. Lett.*, 291: 553–561, 1998.
11. Däubler, T.K., Bittner, R., Meerholz, K., Bittner, R., Meerholz, K., Cimrova, V., and Neher, D., Charge carrier photogeneration, trapping, and space-charge field formation in PVK-based photorefractive materials, *Phys. Rev. B*, 61(20): 13515–13527, 2000.
12. West, P.D., Rahn, M.D., Im, C., and Bässler, H., Hole transport through chromophores in a photorefractive composite based on poly(N-vinylcarbazole), *Chem. Phys. Lett.*, 326(5–6): 407–412, 2000.
13. Shakos, J.D., Rahn, M.D., West, P.D., and Khand, K., Holographic index-contrast prediction in a photorefractive polymer composite based on electric-field-induced birefringence, *J. Opt. Soc. Am. B*, 17(3): 373–380, 2000.
14. Saleh, B.E.A. and Teich, M.C., *Fundamentals of Photonics*, John Wiley & Sons, New York, 1999.
15. Yariv, A., *Quantum Electronics*, 3rd ed., John Wiley & Sons, New York 1989.
16. See chapter 6 of this book.
17. Kogelnik, H., *Bell Syst. Tech. J.*, 48, 2909, 1969.
18. Bant, S.P., Binks, D.J., and West, D.P., Full geometry dependence of index contrast in photorefractive polymer composites, *Appl. Opt.*, 41(11): 2111–2115, 2002.

chapter four

Charge photogeneration

In a photorefractive medium the total fluence of light (the product of the intensity of light and the time of exposure) determines the extent of *progress* towards the final, steady-state response of the material to an optical pattern. The end result, the ultimate response of the material is determined by the relative contrast of the optical pattern and not the intensity. This behavior is rather different to that of most nonlinear optical materials, which usually require a high intensity of light to function well. A photorefractive material can respond to weak patterned illumination if sufficient exposure time is possible to obtain an adequate response.

Photorefraction depends on the optical generation of charge carriers within the medium. These charge carriers must be able to move to a different position for the photorefractive effect to be observed. Experimentally, we can distinguish at least two distinct processes leading to a mobile charge created in response to an optical excitation. These are generation of a bound pair and dissociation of that pair. Initially a photon of light is absorbed at a charge generation site and the energy associated with the light absorbed leads to an excited state, which is stable for a characteristic lifetime. In a suitable material, the excited state may have a high electric dipole moment due to two distinct regions of opposite net charge, a charge transfer (CT) state. If the excitation remains unperturbed, the site may then relax radiatively, releasing a photon of light after a characteristic delay. This is photoluminescence. The site may also relax nonradiatively with energy passing to collective thermal modes within the material. During the lifetime of the excited state there may also be many short-lived partial dissociations of an excited electron from the CT state. Only if a complete dissociation occurs will a free charge carrier and an ion be created. The dissociation process is enhanced strongly by the presence of an electric field to separate the two opposite charges.

Often the first process involved in charge photogeneration in amorphous organic photorefractive media can be considered to be the creation of a pair of opposite charges spaced a characteristic distance apart (the thermalization radius) and bound together by their electric attraction. The efficiency with

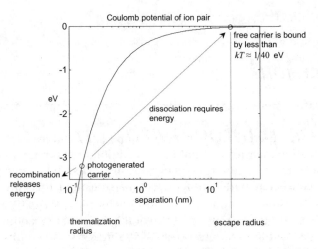

Figure 4.1 The charge generation process in organic media. A pair of ions is created at a distance equal to the thermalization radius. The more mobile of the ions may escape to energy greater than the thermal energy, *kT*, leading to dissociation. Alternatively, after a characteristic lifetime, the two ions may recombine.

which such bound pairs are created from photons absorbed is termed the *primary quantum yield of bound pairs*. The two opposite charges of the bound pair are attracted to each other and this is likely to lead to their recombination, often referred to as *geminate* (that is, twin) *recombination*. Such recombination recreates the neutral site that existed prior to photon absorption.

The bound pair may separate with energy obtained from the thermal environment, or from an applied field. Dissociation may be assumed to occur if the electric potential binding the two charges is reduced by distance to become similar to the mean thermal energy. Once dissociated, a free charge is no more likely to recombine with the original site of photogeneration than any other similarly ionized site (see Figure 4.1).

Ideally in a photorefractive medium one polarity of charge will be much more mobile than the opposite polarity. Either hole or electron transport will dominate the photocurrent in this case. In most amorphous organic media it is hole transport that is most significant. A mobile hole (an electron vacancy) becomes able to move through the material after dissociation, while the negative countercharge remains static and bound to the site of the sensitizer, which is therefore an anion.

The photocurrent observed from a photorefractive material is proportional to the number density of free carriers. Photocurrent measurement allows the determination of the quantum efficiency of free charge generation due to optical absorption. The efficiency of the dissociation of bound pairs may be deduced from observation of the quenching of the photoluminescence from a sample when an electric field is applied to it. To calculate the primary yield, the overall quantum efficiency must be divided by the dissociation efficiency that is assumed to be a contributing factor.

Model of charge generation in organics

The absorption of light is understood to lead to the generation of some pairs of opposite charges. The extent to which these charges are able to avoid geminate recombination will depend on the characteristic distance between them when they are generated, known as the *thermalization radius.*

The mathematics describing the escape from geminate recombination was described for the case of ions in solution by Onsager.[1,2] A classical assumption was used that when the two opposite charges are attracted to the same location their recombination and annihilation is both instantaneous and complete. Even so, amorphous organic media exhibit charge generation properties consistent with Onsager's ion dissociation theory of 1938 if a thermalization radius of a few nanometers is used. Spectroscopic evidence suggests however that only nearest-neighbor interactions are significant between molecules in the charge photogeneration process.[3] Many argue that this limits credible thermalization radii to a few tenths of a nanometer.

A modification used to explain the behavior of charge photogeneration in amorphous organic media is that the pairs of charges that undergo dissociation according to Onsager's theory are the charged regions associated with the first excited state of a molecular charge transfer complex.[4] The Onsager theory of 1934 describes the field-assisted dissociation of the charges well. However, geminate recombination in amorphous organics involves both the reformation of the first excited state, CT_1 and the relaxation of this state back to the relatively unpolar ground state, CT_0, of the CT complex. This relaxation occurs not instantaneously but with a rate determined by the lifetime of the CT_1 state. With this modification, the thermalization radius needed to describe charge generation properly is found to be consistent with nearest-neighbor interactions. During the lifetime of the CT_1 state many partial dissociations may take place due to only a single photoexcitation of the material and this increases the efficiency of dissociation such that the ion pair may be considered bound together much more tightly. Diagrammatically the model of Braun is summarized in Figure 4.2.

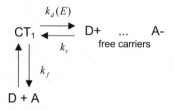

Figure 4.2 The model of Braun (1984) for charge generation in organic media. CT_1 formation is shown due to a donor, D and an acceptor, A. Onsager's 1934 theory of geminate recombination is used to describe the rate of dissociation, $k_d(E)$ from CT_1 to free carriers in the presence of an electric field.

The CT_1 state

The first excited singlet state of a charge transfer complex is characterized by charge separation with two regions of net charge spaced a specific distance apart. The distance is fixed by the molecular structure. Considering these two regions as an ion pair, a Coulomb potential

$$\Delta E_C \approx \frac{e^2}{4\pi\varepsilon_s\varepsilon_0 a} \tag{4.1}$$

can be defined which binds the two charges together. If this binding energy is sufficiently small then the equilibrium between bound and dissociated states will lead to significant numbers of ions dissociated—for forward and reverse rate constants k_d (0) and k_r for dissociation (in the absence of an electric field) and recombination, respectively, according to

$$CT_1 \underset{k_r}{\overset{k_d(0)}{\rightleftharpoons}} D+ \quad \dots \quad A- \tag{4.2}$$

the equilibrium constant for the ion pair

$$K'(0) \approx \frac{k_d(0)}{k_r} \approx \frac{3}{4\pi a^3} \exp\left(\frac{-\Delta E_C}{kT}\right). \tag{4.3}$$

Postulated parameters for an organic system might be an ion pair separation of $a \approx 2 \times 10^{-10}$m, a recombination rate $k_r \approx 2 \times 10^{-12}$m^{3}s^{-1} (from experiments on anthracene crystals) and the dissociation rate in the absence of an electric field is then approximately

$$k_d(0) \approx 10^{15} \exp\left(\frac{-\Delta E_C}{kT}\right). \tag{4.4}$$

Thus, if the potential energy barrier to ion dissociation $\Delta E_C \leq 0.7$ eV, then $k_d \geq 10^3$ dissociations occurring per second. An excited state lifetime of 10 ns would lead in the absence of an electric field to a probability of dissociation of at least 10^{-5} in these circumstances. A lower barrier to dissociation will increase the probability of dissociation according to (4.4).

The Wannier exciton

For the case of a donor (acceptor) matrix with a dilute concentration of acceptor (donor) molecules, the concept of the Wannier exciton provides an alternative theoretical route to charge separation which may in turn lead to dissociation. The equilibrium constant for dissociation of an exciton can be estimated as the product of the Boltzmann factor for exciton promotion to the conduction band and the conduction band density of states

$$K(0) \approx 2 \left(\frac{2\pi m_e kT}{h^2} \right)^{3/2} \exp\left(\frac{-\Delta E}{kT} \right), \tag{4.5}$$

where m_e is the reduced effective mass for the hole (electron) orbiting the electron (hole). With assumptions of an equal effective mass for holes and electrons (both equal to approximately ten times the free electron mass) a dependence on ΔE similar to (4.4) is obtained for the dissociation rate constant $k_d(0)$. The theory of Braun thus may be valid both for ion pair dissociation and for Wannier exciton dissociation.[5]

Field-assisted dissociation

The 1934 Onsager theory of geminate recombination and escape describes the kinetics of dissociation and recombination when an electric field is applied to draw the two opposite charges apart. The dissociation constant in the presence of the field, $K(E)$, is enhanced according to

$$K(E) = K(0) \frac{J_1[2\sqrt{2(-b)^{\frac{1}{2}}}]}{\sqrt{2(-b)^{\frac{1}{2}}}} = K(0) \left\{ 1 + b + \frac{1}{3}b^2 + \frac{1}{18}b^3 + ... \right\}, \tag{4.6}$$

where J_1 is the first Bessel function, the field parameter

$$b = \frac{e^3 E}{8\pi \varepsilon_s \varepsilon_0 k^2 T^2}, \tag{4.7}$$

and the field enhancement does not depend on ion pair separation (thermalization radius).

Langevin recombination

Assuming that the rate of recombination of the geminate pairs is constrained by the process of ion diffusion with low ion mobility, the Langevin form for the recombination rate constant,

$$k_r = \frac{\langle u \rangle e}{\varepsilon_s \varepsilon_0}, \tag{4.8}$$

may be used to obtain estimates of the recombination rate from hole and electron mobilities, averaged spatially, $\langle u \rangle$. When Langevin recombination is assumed the dissociation rate in the presence of an electric field is given by

$$k_d'(E) = \frac{3\langle u \rangle e}{4\pi \varepsilon_s \varepsilon_0 a^3} \exp\left(\frac{-\Delta E}{kT}\right)\left\{1 + b + \frac{1}{3}b^2 + \frac{1}{18}b^3 + ...\right\} \qquad (4.9)$$

in the case of ion pair dissociation, or

$$k_d(E) = \frac{2\langle u \rangle e}{\varepsilon_s \varepsilon_0}\left(\frac{2\pi m_e kT}{h^2}\right)^{\frac{3}{2}} \exp\left(\frac{-\Delta E}{kT}\right)\left\{1 + b + \frac{1}{3}b^2 + \frac{1}{18}b^3 + ...\right\} \qquad (4.10)$$

for the dissociation of excitons. Thus, a higher mobility increases the dissociation rate and a longer excited state lifetime yields a greater number of free charges.

A special case: PVK and TNF

The donor matrix PVK is photoconductive at visible wavelengths when combined with TNF. The trinitrofluorenone is able to associate closely with the carbazole side chain of the PVK. The structures of the repeat unit of PVK and of TNF are shown in Figure 4.3. These structures are very similar in shape, both are planar with a delocalization of π-electrons above and below the σ-bonded molecular plane. The TNF contains additional nitro groups that withdraw electron density from the delocalized system in the excited state.

The PVK:TNF charge transfer system has been studied closely. [6] When in close proximity to each other, the electronic systems of the molecule and the side chain unit can become coupled, leading to correlated collective electronic behavior as an interactive quantum well. Electron density in this complex is transferred partially from the carbazole unit of the PVK side chain to the TNF molecule. In the ground state only about 20% of an electronic charge is transferred on average, but new optical absorption is observed due to the correlated electron systems of the complex (Figure 4.4).

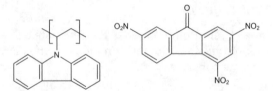

Figure 4.3 Structures of poly(N-vinylcarbazole), PVK (left) and 2,4,7-trinitro-9-fluorenone, TNF (right).

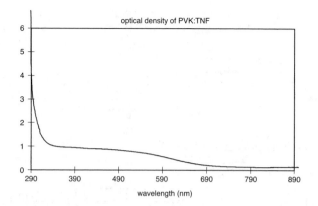

Figure 4.4 The optical density (absorbance) of light in a solid composite of PVK:TNF consisting of 90 %wt. PVK. A broad shoulder extends across the visible region of the spectrum that is not observed with either material in pure form. This shoulder is due to PVK:TNF charge transfer complex formation.

The energy of direct excitation from the highest occupied molecular orbital (HOMO) of the complex to the lowest unoccupied orbital (LUMO) is 2.17 eV and the absorption line width for this transition is 0.50 eV at a temperature of 295 K. This corresponds to a spectral absorption peak at 570 nm with width in excess of 100 nm. If the first excited singlet state is populated then the charge transfer to the TNF from the PVK repeat unit reaches 80% of an electronic charge on average. The first excited state, CT_1, corresponds to something approaching a positively ionized carbazole with a negatively ionized TNF molecule, spaced apart by the 0.15 nm interplanar separation in the complex. The transition from ground to excited states is associated with a dipole moment increase of 0.85×10^{-10} e m (about 5 Debyes) and an increase in the polarizability of 4.2×10^{-21} e m^2/V.

The approximation that the excited state of the complex corresponds to the creation of an ion pair is illustrated in the energy level diagram of Figure 4.5. The resonance peaks at a wavelength of about 570 nm. A relaxation rate of the excited state of $k_f \approx 3 \times 10^9$ s^{-1} can be used with a thermalization radius $a = 1.5 \times 10^{-10}$ m.[7] The overall quantum efficiency of charge photogeneration can rise by a factor 10 from 10^{-5} to 10^{-4} when the field is doubled from 20 to 40 V μm^{-1} in this model.

Some information can be obtained about the sequence of events involved in the photogeneration of free charge in PVK:TNF from fast optical pulse excitation of the material. When an intensity pattern is formed using two interfering optical pulses of duration 200 fs, the excitation leads to a short-lived hologram. The hologram can be probed with another fast pulse and the amplitude of the hologram is found to decay away with a nonexponential transient after the writing pulse has passed. In contrast to this, two coherent light waves with perpendicular linear polarizations overlapped within the material will form an optical polarization state that varies with

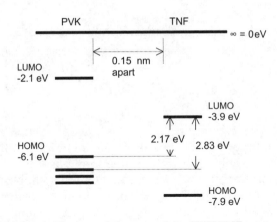

Figure 4.5 Energy level diagram of the composite PVK:TNF, pictured as ion pairs. In this picture the creation of the CT_1 state corresponds to electron transfer from the highest occupied orbital (HOMO) of PVK to the lowest unoccupied orbital (LUMO) of TNF.

location in the material. Here an orientational pattern of the material excitation results, which decays with a simple exponential transient.

These results can be explained using a dynamic model developed from the general case. A charge-separated state is added to the rate model as an intermediate between the CT_1 excited state and the dissociated charges, as shown in Figure 4.6. It is tempting to imagine the charge-separated state to correspond to a weak nearest-neighbor interaction between an anionic complex of PVK:TNF and an adjacent cation of a carbazole side chain on the PVK macromolecule, one repeat unit along the polymer backbone. If so, this would correspond to the situation after a single hop of a hole from the CT_1 complex state to an adjacent carbazole (an electron transfer from a neighboring carbazole onto the CT_1 state). PVK forms a donor matrix and subsequent hole transport would free the two opposite charges from their binding Coulomb potential as screening reduces the binding energy with increased

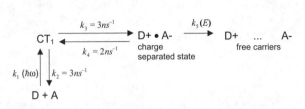

Figure 4.6 The photogeneration model of Walsh and Burland for PVK:TNF (1:0.4 wt.). After excitation of the CT_1 state, a transition to a charge-separated state occurs with rate coefficient $k_3 = 3$ ns^{-1}. This separated state may either dissociate, with a field-dependent rate coefficient $k_5(E)$, or it may relax. Relaxation may be back to CT_1 or it may be back to the neutral ground state; the experiment did not resolve this. The rate coefficients shown were found to fit to experimental data.

separation to the mean thermal energy, kT. This specific developed model may therefore represent an allowance for the discrete molecular nature of the charge transport process in amorphous organic media, in contrast to the transport conduction bands of crystals.

Exciton diffusion as a route to CT₁

The formation of a Wannier exciton (or a Frenkel exciton that is bound together more tightly) forms a possible alternative route to a CT_1 state of excitation for charge photogeneration in amorphous organics. The idea is that optical absorption occurs within a material that supports exciton formation.[8] The exciton is a distortion of the electronic states that persists long enough to allow diffusion of the energy contained within it. A bipolaronic exciton is neutral overall, containing both a positive and an equal negative charge so it does not drift in an electric field. The exciton may diffuse to the site of a dopant with a different electron affinity, at which a CT_1 state may be formed between donor and acceptor as the exciton decays. An exciton-derived photogeneration rate model is shown in Figure 4.7.

The key difference between the exciton model of Goliber and Perlstein and the model of Braun is the manner in which the CT_1 state is formed. These authors advocated one other refinement to the Onsager picture of the geminate recombination process. The original theory of Onsager was to describe the process of ion dissociation in liquids. The solid amorphous environment of an organic photorefractive material will be inhomogeneous locally and an improved description of the behavior of the material will be obtained if the thermalization radius is assumed to have a Gaussian probability distribution. Braun's theory had assumed a single-valued thermalization radius, for simplicity.

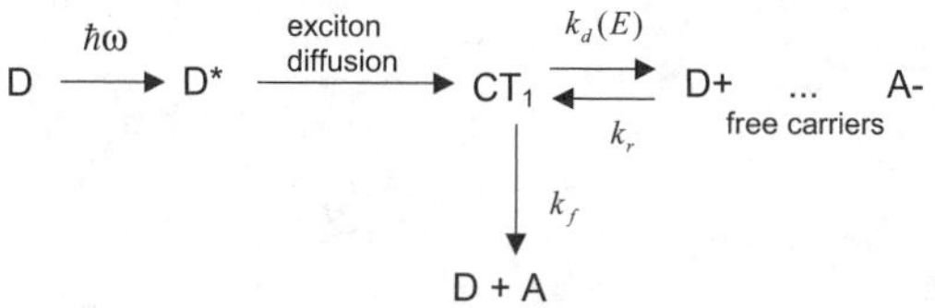

Figure 4.7 The model of Goliber and Perlstein, adapted from Braun. In this, a donor matrix is capable of exciton formation. A photogenerated electronic disturbance (exciton) propagates only by diffusion through the donor matrix. When the exciton meets an acceptor site, a CT_1 state may be formed as the exciton decays. The field-assisted dissociation of the CT_1 state is then described by the Onsager 1934 theory, as it is in Braun's model. The exciton diffusion model was found to work for films of bisphenol-A-polycarbonate (Lexan 145) doped with 40 %wt. triphenylamine (TPA) and small amounts of the acceptor difluoroboron-1,3-diphenyl-1,3-propanedionate (DPBDK).

Most amorphous organic media for photorefraction use bimolecular CT_1 formation from direct photoexcitation. Why is exciton diffusion potentially an attractive alternative? Exciton formation within a suitable material may be either a broad band or a more sharply defined process spectroscopically. If specific resonances are selected by an appropriate choice of wavelength then it may be possible to maximize the efficiency of the photogeneration process. Excitons must diffuse to the edge of the exciton-supporting medium. This process is efficient if the exciton diffusion length,

$$l = \sqrt{D\tau_{exc}} \ , \tag{4.11}$$

is longer than the distance from the exciton formation site to the interface. Here, D is the exciton diffusion coefficient and τ_{exc} is the exciton lifetime.

Fick's law for the diffusion of exciton density, $n(x)$, in a one-dimensional region of thickness L states that the rate of change of exciton density

$$\frac{\partial n(x)}{\partial t} = \alpha I_0 \exp(-\alpha x) + D \frac{\partial^2 n(x)}{\partial x^2} - \frac{n(x)}{\tau_{exc}} \tag{4.12}$$

and in the simple case of boundary conditions $n(0) = n(L) = 0$ the solution for exciton density in the steady state is

$$n(x) = \frac{I_0 \alpha l^2}{D(1 - \alpha^2 l^2)} \left\{ e^{-\alpha x} + \frac{e^{x/l}\left(e^{-L/l} - e^{-\alpha L}\right)}{e^{L/l} - e^{-L/l}} - \frac{e^{-x/l}\left(e^{-L/l} - e^{-\alpha L}\right)}{e^{L/l} - e^{-L/l}} \right\} . \tag{4.13}$$

The efficiency, η, of generation of the CT_1 states through rapid exciton decay at the boundaries

$$\eta \propto \frac{J(0) + J(L)}{I_{abs}} \tag{4.14}$$

where the fluxes of excitons are

$$J(0, L) = D \frac{\partial n(x)}{\partial x} \bigg|_{x=0,L} \tag{4.15}$$

and the intensity absorbed in the material sustaining the excitons is

$$I_{abs} = I_0 \left(1 - \exp(-\alpha L)\right) .$$

This analysis, after Mulder,[9] yields a sensible result for a thick sample of excitonic material. Using the limits of a thick materal ($L \gg l$) and weak absorption ($\alpha^{-1} \gg l$) the rate of photogeneration of CT_1 states is proportional to the product, $\alpha l I_{abs}$, of absorption coefficient, intensity absorbed and diffusion length.

A different situation arises if the excitonic material can be made small relative to the exciton diffusion length, $L \ll l$. Using again the limit of weak absorption ($\alpha^{-1} \gg l$), the rate of photogeneration of CT_1 states is proportional to the product $\alpha^2 l^2 I_{abs}$, now quadratic in absorption coefficient. This can be interpreted as a linear dependence of intensity gradient on absorption coefficient and also a linear dependence of exciton flux on the number density gradient, combining in the absence of exciton losses during diffusion to yield superlinear increases in CT_1 yield with increased absorption. This situation will be difficult to achieve in practice because exciton diffusion lengths in amorphous organic media are typically of the order of a few ×10 nm.

Laminar photoreceptors have long been used in the xerographic industry. These layered devices alternate donor and acceptor material. The rate of CT_1 generation in them has been found to be independent of electric field due to an interfacial enhancement of the exciton decay process. Marcus theory[10] argued that electron transfer at an interface between donor and acceptor media would involve reorganization of the states of those molecules. The energetic barrier to this would be minimized if the first singlet excited states of the donor and acceptor lie close together in energy, such that a polarization in one may influence the other strongly. The efficiency of electron transfer will then depend on the Franck-Condon coefficients.

The interface will enhance the CT_1 formation rate significantly with a material combination where the energy released by the reorganization of two neutral molecules into a pair of ions matches closely the energy required for electron transfer. The yield of CT_1 will not depend on electric field. The process of CT_1 dissociation into free charges is dependent on field strongly. The dissociation efficiency should be greatest when the spatial average of electron and hole mobility away from the CT_1 site is high. In a photorefractive material it is desirable that one polarity of charge carrier is much more mobile than the other polarity, to avoid competition between holographic gratings of differing formation rates. It is also desirable to have a material that is homogeneous on the scale of the wavelength of light, around 0.5 μm, in order to avoid optical scatter. If exciton-supporting acceptor particles significantly smaller than optical wavelengths could be used within a donor matrix then the area of the interface between donor and acceptor could be very large and optical scatter may not be too high. Exciton diffusion to the interfaces could be efficient for very small particle sizes. If the surface enhancement effect is strong then the yield of CT_1 states could be relatively strong and independent of applied field. An acceptor with a fair electron mobility might improve the dissociation efficiency while electron transport confined within small particles may avoid competition effects in a photorefractive material.[11]

Charge generation field dependence affects photorefractive contrast, spatial phase and rate of response

The dissociation of charges is assisted strongly by the application of an electric field in amorphous organic materials. This fact was not accounted for in the standard theory of photorefraction outlined in much of chapter 3, in which the photogeneration rate was assumed to be a single-valued material property. (This assumption was based on the properties of photorefractive crystals in which the crystal field dominates the charge generation process.) What will be the effect on a photorefractive hologram when a field-dependence of charge photogeneration efficiency is introduced?

In most cases a photorefractive hologram is not trap density limited in its contrast. A space charge field forms with final amplitude equal to the projection of the poling field along the grating vector (Figure 4.8). The size of the resultant field

$$E_{res}^2 = (E_0')^2 + E_{sc}^2 \sin^2 Kx + 2E_0' E_{sc} \cos\phi \sin Kx \qquad (4.16)$$

where $E_{sc} \leq E_0' \cos\phi$ and through the hologram the electric field is patterned in amplitude such that the resultant field squared is bounded by

$$(E_0')^2 (1 - \cos^2\phi) \leq E_{res}^2 \leq (E_0')^2 (1 + 3\cos^2\phi) . \qquad (4.17)$$

The electric field variation depends on the orientation of the grating vector, $\underline{K}$, relative to the poling field, E_0' . In a typical configuration the projection angle ϕ is nearly $60°$ and $\dfrac{3}{4}(E_0')^2 \leq E_{res}^2 \leq \dfrac{7}{4}(E_0')^2$. From (3.31) away from saturation the phase of the space-charge field modulation is nearly opposite to the phase of the optical pattern creating it, as in Figure 4.9. Often even high poling fields do not move this grating more than $20°$ from antiphase because in many materials of interest, the saturation field, E_q, is 50 Vμm^{-1} or more and the phase shift is approximately

$$\phi \approx \arctan\left(\frac{-E_0}{E_q}\right) \qquad (4.18)$$

when the diffusion field can be neglected.

The effect of the electric field patterning may be to enhance the photogeneration rate in the darker regions of the optical pattern, as long as the modulation in the electric field is not too great. This occurs at the expense of reduced photogeneration efficiency in brighter regions. This argument

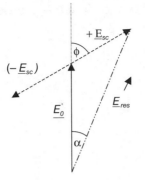

Figure 4.8 Orientation angles of the poling field, E_0' and the space-charge field, $\pm E_{sc}$, within a photorefractive hologram (after Figure 3.10).

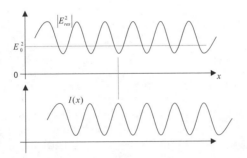

Figure 4.9 Electric field and optical intensity patterns are displaced and typically almost out of phase in a photorefractive hologram. (Top) The resultant electric field and (bottom) the optical intensity pattern that caused it.

suggests that field-dependent charge generation efficiency may lead therefore to a reduction in the contrast of the hologram.

There may be a significant drop in the quantum efficiency of charge generation for a hologram close to the steady state, when the recording process is nearly complete. Initially, prior to hologram recording, there is no space-charge field and the resultant electric field is uniform throughout the material. A holographic recording transient will not be affected by field-dependent photogeneration early in the recording, but the hologram will saturate earlier and at a lower contrast than for a comparable material with a photogeneration efficiency independent of the applied field, as shown in Figure 4.10.

The effect of field dependence of photogeneration on photorefraction may be modeled analytically where a low contrast allows the linearization of the material equations.[12] A useful expression for the field dependence of the photogeneration coefficient, s, is that at any particular field strength, E, the field dependence of photogeneration may be described by $s \propto E^p$. This expression is valid as long as the pattern in the magnitude of the electric

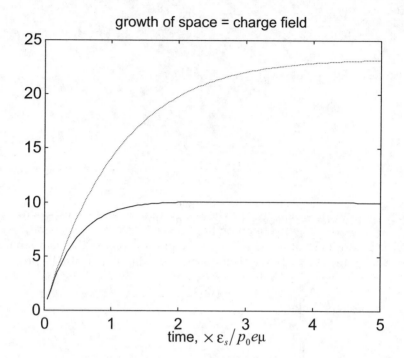

Figure 4.10 The effect of field dependence of photogeneration efficiency. The solid line shows the growth of the space charge field (V μm⁻¹) under an applied field of 50 V μm⁻¹ at an angle of 60° to the grating vector. The Braun model has been used and the photogeneration efficiency has been assumed to be much less than unity. The other parameters used were E_q = 125 V μm⁻¹, E_M = 50 V μm⁻¹ and E_D = 0.05 V μm⁻¹, m = 0.95. The dotted line shows the corresponding growth under the assumption of a photogeneration efficiency independent of field. Field dependence has lessened the contrast of the hologram and accelerated the approach to the steady state. Perhaps surprisingly, this effect will diminish if the photogeneration quantum efficiency can be increased nearer to unity.

field is not too great. The linearized, coupled equations for anion density, field and hole density must be revised to account for the patterning of the photogeneration rate and equation 3.17 becomes

$$\frac{\partial}{\partial t}\tilde{N}_A^-\bigg|_{\Delta\tilde{s}\neq0} \approx (s\Delta\tilde{I}+I_0\Delta\tilde{s})(N_A-N_D^+)-sI_0\Delta\tilde{N}_A^- - \gamma_R(p_0\Delta\tilde{N}_A^-+\Delta\tilde{p}N_{A0}^-) \quad.(4.19)$$

The extra term in $\Delta\tilde{s}$ is due to the patterning of the photogeneration efficiency by the space-charge field. This leads to a modification in the coefficient, B, of the differential equation 3.20 for the space-charge field

$$A\frac{\partial}{\partial t}\Delta\tilde{E} + B\Delta\tilde{E} = jC \tag{4.20}$$

that for a field-dependent photogeneration efficiency

$$B\Big|_{\Delta\tilde{s}\neq 0} = 1 + p\frac{E_0}{E} + \frac{E_D + j\left[E_0 - \left(\dfrac{pE_D E_q}{E}\right)\right]}{E_q} \tag{4.21}$$

where two new terms dependent on p have appeared, dependent on the value, E, of the local field strength before patterning and the exponent, p, of the power law for photogeneration field dependence. In the general case of applied poling field at an angle, ϕ, to the grating vector, the field strength along the grating $E_0 = E_0' \cos\phi$. The space-charge field must also be projected back along the axis of the poling field to obtain the impact of the space-charge field on the local field determining the photogeneration efficiency. For the case of nonparallel fields

$$B\Big|_{\Delta\tilde{s}\neq 0} = 1 + p\cos^2\phi + \frac{E_D + j\left[E_0 - \left(p\cos^2\phi\dfrac{E_D E_q}{E}\right)\right]}{E_q}. \tag{4.22}$$

In typical conditions for amorphous photorefractive materials the diffusion field is neglected $E_D \ll E_0, E_q$ and the saturated electric field in this case

$$\left|\Delta\tilde{E}\right|_{t\to\infty} \approx \left|\frac{-E_q E_0}{\left([1 + p\cos^2\phi]E_q + jE_0\right)}\frac{\Delta\tilde{I}}{I_0}\right| \tag{4.23}$$

while the phase of the hologram under similar assumptions is modified accordingly to become for the case of field-dependent photogeneration

$$\Phi = \arctan\left(\frac{imag(jC/B)}{real(jC/B)}\right) = \arctan\left(\frac{-E_0}{[1 + p\cos^2\phi]E_q}\right). \tag{4.24}$$

Finally, the corresponding complex rise time of approach toward the steady-state space-charge field is now

$$\tau_{rise} = \frac{A}{B} = \frac{\varepsilon_s}{p_0 e \mu} \frac{(E_M + jE_0)}{E_M} \frac{E_q}{([1 + p\cos^2\phi]E_q + jE_0)} \, . \tag{4.25}$$

The above results incorporate the exponent p that has arisen from the theory of photorefraction with field-dependent photogeneration efficiency. The exponent arises in the photorefraction theory from the calculation of

$$\frac{\Delta\tilde{s}}{\Delta\tilde{E}} \frac{E_0}{s} = p \tag{4.26}$$

following naturally from the simple relation

$$s \propto E^p \, . \tag{4.27}$$

It is possible to calculate the value of the exponent for any specific field without the need for material properties. The photogeneration coefficient will be proportional to the probability of dissociation

$$P(E) = \frac{k_d(E)}{k_f + k_d(E)} \tag{4.28}$$

and therefore the field dependence of this is given by

$$\frac{\partial}{\partial E} P(E) = [1 - P(E)]P(E)\frac{1}{k_d}\frac{\partial k_d}{\partial E} \, . \tag{4.29}$$

Using the Onsager field dependence of ion dissociation leads to the result

$$k_d \propto 2\frac{J_1(jx)}{jx} \tag{4.30}$$

where $J_1(jx) = I_1(x)j$ is the first modified Bessel function and $x = \sqrt{8b}$ where

$$b = \frac{e^3 E}{8\pi\varepsilon_s\varepsilon_0(kT)^2} \, . \tag{4.31}$$

Differentiating the dissociation rate constant

$$\frac{\partial k_d}{\partial x} = \frac{I_2(x)}{I_1(x)}k_d \tag{4.32}$$

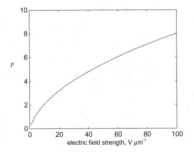

Figure 4.11 The electric field dependence of the exponent, p, used to describe the field dependence of the photogeneration efficiency according to $s \propto E^p$. The calculation $p \approx x\, I_2(x)/2I_1\,(x)$ is valid generally for dissociation from an organic excited state in the approximation of quantum efficiency much less than unity.

leading to the result that

$$p = \frac{\Delta \tilde{s}}{\Delta \tilde{E}} \frac{E_0}{s} = [1 - P(E)] \frac{I_2(x)}{I_1(x)} \frac{x}{2} \tag{4.33}$$

and in the approximation $P(E) \ll 1$, the field dependence exponent is plotted in Figure 4.11. If a material can be found for which the quantum efficiency of free carrier dissociation, $P(E) \to 1$ then the value of p and the effect of field dependence on hologram formation will be reduced relative to the values given in this figure.

Choice of photosensitizer

Much of the early work on photorefractive polymers and organic glasses was based on the prior understanding of photorefraction in crystals. Within organic crystals the optical generation of charge carriers is associated usually with the oxidation of defect or impurity sites within the optical band gap of the crystal, such as iron impurity in lithium niobate and other ferroelectric crystals. Host matrices were adopted for polymer composites and organic glasses that, when pure, are transparent to visible light. The wide band gap between the highest occupied and lowest unoccupied molecular orbitals (HOMO and LUMOs) corresponds typically to materials with optical resonances in the near ultraviolet region. Most organic materials form hole transporters more efficiently than they form electron transport media. Charge generation with visible light is due usually to the addition of small concentrations of a dopant with lower HOMO and LUMO levels than the host matrix. For such an addition of an acceptor into a donor matrix it is the LUMO of the dopant which must be deeper in energy than the LUMO of the host and within visible photon range (2 eV) of the HOMO of the host.

Most of the potential applications for photorefractive materials are based on the ability to create holograms within the bulk of the material. For a homogeneous medium with low optical scatter, it is usual to dope the matrix with a low concentration (e.g., 1%wt.) of sensitizer to produce a weak optical absorption ($\alpha \leq 50$ cm^{-1}) at a typical operating wavelength in the range 650 nm $\leq \lambda \leq 750$ nm. The primary quantum yield of CT_1 states may be as high as 40%, although relatively few reports have been made of this. The field-assisted dissociation of these states leads to a quantum efficiency of charge photogeneration reaching a few percent at high electric fields up to 100 V μm^{-1}.

A simple model of the charge generation limit

A patterned illumination intensity

$$I = I_0 \left(1 + m \cos Kz\right) \tag{4.34}$$

will lead to a patterned photogeneration of holes

$$g(z) = \frac{\phi \alpha I(z)}{\hbar \omega} \tag{4.35}$$

where ϕ is the quantum efficiency for photogeneration, α is the absorption coefficient for light in the medium and $\hbar \omega$ is the photon energy. The excess holes generated in the brighter regions move into the darker regions as shown in Figure 4.12. The pattern of holes created per grating period $\Lambda = 2\pi/K$, per unit time is

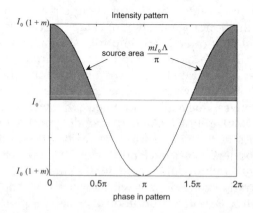

Figure 4.12 The regions of above average intensity become net sources of hole density. Excess hole density generated here will recombine in regions of below average intensity. This patterns the distribution of immobile ions.

$$\int_{-\Lambda/4}^{\Lambda/4} \frac{\phi}{\hbar\omega} \alpha m I_0 \partial z \cos Kz = \frac{\phi}{\hbar\omega} \alpha m I_0 \frac{\Lambda}{\pi} . \tag{4.36}$$

Let transport and trapping processes be idealized such that each photogenerated charge is trapped instantaneously at a trap site within a dark region. A nonuniform hole distribution

$$p(z) = p_0 + \Delta p \cos(Kz + \xi) \tag{4.37}$$

builds up, displaces a distance ξ radians through the optical pattern. Within one period of the pattern, the total quantity of charge moved is

$$\int_{p > p_0} e\Delta p \partial z \cos(Kz + \xi) = e\Delta p \frac{\Lambda}{\pi} . \tag{4.38}$$

The total charge moved will clearly be less than or equal to the total pattern of holes created by the optical pattern, so

$$\frac{\partial}{\partial t} \Delta p \leq \frac{\phi}{\hbar\omega} \alpha m I_0 . \tag{4.39}$$

Using Poisson's rule that the divergence of the electric field

$$\varepsilon_s \frac{\partial}{\partial z} \Delta E = e\Delta p \cos(Kz + \xi) \tag{4.40}$$

depends on the net charge density, the rate of growth of the space charge field is constrained by the need for charge generation to the limit

$$\frac{\partial}{\partial t} \Delta E \leq e \frac{\phi}{\hbar\omega} \frac{\alpha m I_0}{\varepsilon_s K} . \tag{4.41}$$

Using accessible values for the parameters, that $e = 1.6 \times 10^{-19}$ C, the grating vector $K = 2\pi/(4 \times 10^{-6})$ m^{-1}, the static dielectric constant $\varepsilon_s = 3 \times 8.85 \times 10^{-12}$ F m^{-1}, the intensity pattern contrast $mI_0 = 200$ mW cm^{-2}, the absorption $\alpha = 10$ cm^{-1}, the photon energy $\hbar\omega = 1.7$ eV and the quantum efficiency $\phi = 10^{-2}$, the space-charge field will build up with a rate limited by the need for charge generation to

$$\frac{\partial}{\partial t} \Delta E \leq 0.3 \quad \text{V } \mu\text{m}^{-1} \text{ msec}^{-1}.$$

Effect of a finite mobility on the charge generation limit

In chapter 3 the standard theory of photorefraction was used with a single-valued mobility to obtain the limit to the growth rate of space-charge field (3.30), repeated here, that when diffusion is negligible

$$\left| \frac{\partial}{\partial t} \Delta \tilde{E} \right| \leq \frac{p_0 e \mu}{\varepsilon_s} \left| \frac{-E_M E_0}{(E_M + jE_0)} \frac{\Delta \tilde{I}}{I_0} \right| \tag{4.42}$$

taking into account hole trapping as recombination of holes with anions. Substituting for the saturated hole density

$$p_0 = sI_0(N_A - N_D^+)\tau_h = \alpha I_0 \frac{\phi}{\hbar \omega} \tau_h \tag{4.43}$$

and the pattern contrast

$$m = \left| \frac{\Delta \tilde{I}}{I_0} \right| \tag{4.44}$$

and the mobility field parameter

$$E_M = (\mu \tau_h K)^{-1} \tag{4.45}$$

we obtain

$$\frac{\partial}{\partial t} \Delta E \leq e \frac{\phi}{\hbar \omega} \frac{\alpha m I_0}{\varepsilon_s K} \left(\frac{-E_0}{E_M + jE_0} \right) \tag{4.46}$$

where the last factor represents the effect of sub-infinite mobility. For a finite mobility the charge generation limit is constrained by the new sub-unity amplitude factor $E_0 / \sqrt{E_M^2 + E_0^2}$. The mobility will have a significant effect on the rate of field growth if the mobility field (the field necessary for a hole to drift one radian in a lifetime) becomes greater than or equal to the applied poling field projection along the grating vector, E_0.

Although relatively little is known about the hole lifetime in photorefractive composites, one estimate has been $\tau_h \approx 0.5$ msec and in a typical poling field of 50 V μm^{-1} with a grating period $\Lambda \approx 3$ μm, the mobility will only restrain space-charge field formation if the mobility is lower than $\mu \leq 2 \times 10^{-7}$ cm^2 (Vs)$^{-1}$.

Pre-illumination effects

Uniform and intense illumination of a photorefractive material for longer than the hole lifetime will ensure that the carrier population is saturated prior to recording a hologram. If it is true that hole traps are neutral sites initially, before a uniform ionized sensitizer population is built up, then uniform pre-illumination will establish the uniform distribution of sensitizer ions which is assumed in the single-trap model of photorefractivity. The charged trap sites will be more effective traps than the neutral sites. For both of these reasons the transients associated with photocurrents and the photorefractive response can be accelerated by a few seconds of uniform pre-illumination.

If the pre-illumination persists for a much longer time in some composites, the rate of response of the material in photoconductivity experiments can be accelerated still further. There will be a reduction in the magnitude of the photocurrent in this case.[13] The extended pre-illumination may be concluded to have led to an increase in the number of sensitizer ions, N_A^-, with a corresponding increase in the number of (passive) compensator ions, N_D^+. In the case of C_{60} as a sensitizer such an increase in the population density of the anion has been confirmed spectroscopically.[14] Hole recombination becomes, therefore, more likely after extended pre-illumination and the photocurrent transient is shortened due to the reduced hole lifetime. The photocurrent magnitude is dependent on the product of the generation rate and the lifetime of holes and this also drops after extended pre-illumination in such materials.

The charge generation limit to hologram formation may depend on the hole lifetime if the mobility is small because of the mobility-field dependence of the charge generation limit when $E_M \geq E_0$. A smaller hole lifetime will increase the mobility field and this may slow down the space charge field formation rate, or it may not affect it where the mobility is sufficient. The rise time for the holographic field depends on the dielectric relaxation time, $\varepsilon_s/p_0 e\mu$. While both the mobility field, E_M and the saturation field, E_q, are nearly proportional to the hole lifetime, the main effect of reduced lifetime on hologram recording rate will be through the fall in the saturated hole density, p_0. This will increase the holographic contrast rise time. Ultimately higher holographic contrast may be observed after extended pre-illumination due to the increased trap density.

References

1. Onsager, L., *J. Chem. Phys.*, 2: 599, 1934.
2. Onsager, L., Initial recombination of ions, *Phys. Rev.*, 54: 554–557, 1938.
3. Merski, J. and Eckhardt, C.J., Piezomodulation spectroscopy of molecular crystals, parts 1–4, *Chem. Phys.*, 75(8): 3691–3742, 1981.

4. Braun, C.L., Electric-field assisted dissociation of charge-transfer states as a mechanism of photocarrier production, *J. Chem. Phys.*, 80(9): 4157–4161, 1984; Smirnov, S.N. and Braun, C.L., Photoinduced free carrier yields in high fields: Field dependent distribution functions, *J. Imaging Sci. Tech.*, 43(5): 425–429, 1984.

5. Morris, R. and Silver, M., Direct electron-hole recombination in anthracene, *J. Chem. Phys.*, 50: 2969, 1969.

6. Weiser, G., Absorption and electroabsorption on amorphous films of polyvinylcarbazole and trinitrofluorenone, *Phys. Stat. Sol. A*, 18: 347–359, 1973.

7. Walsh, C.A. and Burland, D.M., Picosecond photoionization and geminate recombination in an organic donor-acceptor complex, *Chem. Phys. Lett.*, 195(4): 309–315, 1992.

8. Goliber, T.E. and Perlstein, J.H., Analysis of photogeneration in a doped polymer system in terms of a kinetic model for electric field assisted dissociation of charge transfer states, *J. Chem. Phys.*, 80(9): 4162–4167, 1984.

9. Mulder, B.J., Diffusion and surface reactions of singlet oxygen in anthracene, *Philips Res. Rept. Suppl.*, 4: 1, 1968.

10. Marcus, R.A., On the theory of oxidation-reduction reactions involving electron transfer. V. Comparison and properties of electrochemical and chemical rate constants, *J. Chem. Phys.*, 67: 853–857, 1963.

11. Binks, D.J., Bant, S.P., West, D.P., O'Brien, P., and Malik, M.A., CdSe/CdS core/shell quantum dots as sensitizer of a photorefractive polymer composite, *J. Mod. Opt.*, 50(2): 299–310, 2003.

12. Khand, K., Binks, D.J., and West, D.P., Effect of field-dependent photogeneration on holographic contrast in photorefractive polymers, *J Appl. Phys.*, 89(5): 2516–2519, 2001.

13. Hendrickx, E., Zhang, Y.D., Ferrio, K.B., Herlocker, J.A., Anderson, J., Armstrong, N.R., Mach, E.A., Persoons, A.P., Peyghambarian, N., and Kippelen, B., Photoconductive properties of PVK-based photorefractive polymer composites doped with fluorinated styrene chromophores, *J. Mater. Chem.*, 9: 2251–2258, 1999.

14. Grunnet-Jepson, A., Wright, D., Smith, B., Bratcher, M.S., DeClue, M.S., Siegel, J.S., and Moerner, W.E., Spectroscopic determination of trap density in C_{60} sensitized photorefractive polymers, *Chem. Phys. Lett.*, 291: 553–561, 1998.

chapter five

Charge transport in amorphous photorefractive media

The mobility of charges is important in the standard model of photorefraction. If the charge mobility is sufficient for charges to drift one radian through the intensity pattern within one lifetime due to the poling field, then the charge generation limit to the rate of growth of space-charge field is reached. If the mobility of the charge in the medium is lower than this, then the process of charge transport will slow down the photorefractive effect.

Time-of-flight measurements of mobility

Charge transport occurs in a partially filled conduction band of excited energy levels in the crystals for which the standard photorefraction model was developed. The average drift velocity of a free charge distribution in a crystal is proportional to the electric field that induces the drift motion. The constant of proportionality is the mobility, μ, in units typically of cm^2 (Vs)$^{-1}$.

The simplest way to measure mobility is the time-of-flight (TOF) experiment, as shown in Figure 5.1. A flash of light is used to photogenerate a thin layer of free charges next to an electrode of similar polarity (such as holes next to an anode). The electric field in the sample makes the sheet of charges drift toward the opposite electrode. As the sheet of charges moves through the medium the electric potential of the electrodes is changed and a current must flow in the external circuit to maintain the sample at a constant voltage. Where diffusion can be neglected, the current flowing in the circuit will be a step function as in Figure 5.2a.

Diffusion makes the charges develop a Gaussian distribution of arrival times. The width of the distribution increases as $\sqrt{t}$. The tail of the photocurrent in this case has a t^{-2} dependence on time as in Figure 5.2b. The length of the photocurrent transient thus reveals the transit time for charges across the sample of thickness d from

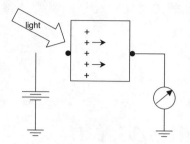

Figure 5.1 The linear time-of-flight experiment (TOF). In this case, a flash of light creates a thin sheet of photogenerated holes next to the anode. These drift in a field toward a cathode. The photocurrent from the cathode is studied as the charges move through the test medium.

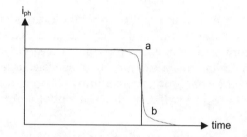

Figure 5.2 Idealized photocurrent vs. time for a crystal in a TOF experiment. (a) No diffusion occurs and a thin charge sheet drifts to the opposite electrode. When the charges reach the electrode, the photoconductor no longer conducts and the current stops. (b) Charges diffuse (dashed line) such that the charge sheet spreads out into a Gaussian distribution before arrival at the opposite electrode.

$$t_{tr} = \frac{d}{\mu E} \qquad (5.1)$$

while the width of the Gaussian distribution of arrival times of charges is quite distinct from this, dependent on the diffusion coefficient, D, and the sample thickness as

$$\sigma_{arrival} \propto dD\sqrt{t} . \qquad (5.2)$$

Such behavior as described above is not observed, however, when performing TOF experiments in amorphous photorefractive composites. Attempts to calculate a diffusion coefficient from TOF data on these materials do not produce a single-valued material parameter in the same way as for crystals. In practice the diffusion coefficient obtained depends on parameters of the measurement process such as the thickness of the sample. The problem

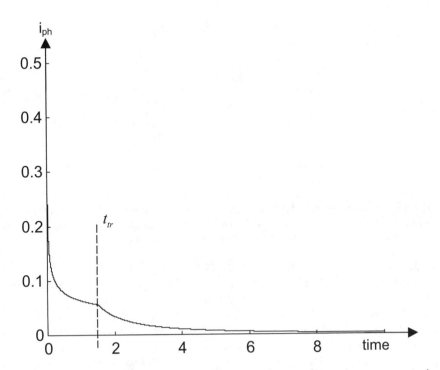

Figure 5.3 A TOF photocurrent transient that indicates dispersive transport. An initial peak of photocurrent falls to a small plateau. This current then falls away again after the transit time, t_{tr}, followed by a long-lasting tail of weak, delayed photocurrent.

in the data analysis is that photocurrent transients from photorefractive composites are closer in form to Figure 5.3 than to Figure 5.2. An initially high photocurrent might fall to a short-lived plateau value before falling again (at the transit time, t_{tr}) in a slow decay that is not exponential. The idea of a diffusion coefficient does not work with this kind of data. When studying a particular material, all the photocurrent transients for different sample thickness or different applied field often look the same if normalized by their values at the end of the plateau. This is known as universality. The shape of the long tail of the distribution is not independent from the transit time, so that a diffusion coefficient cannot be isolated and identified.

Clearly the charges are spreading far apart during their transit across a photorefractive composite in this manner. Such transport is called dispersive transport for this reason.

The shape of the photocurrent transient may be easier to identify on a double-logarithmic plot as in Figure 5.4. A crystalline (mobility dispersion-free) photocurrent transient would be flat until close to the transit time, as charge travels across the crystal. The tail of the photocurrent transient would have a slope of –2 as the last charges reach the target electrode, due to diffusion. The photocurrent transient from an amorphous photorefractive

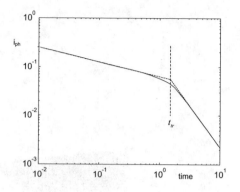

Figure 5.4 A double-logarithmic plot of the TOF dispersive photocurrent transient shown in Figure 5.3. The slopes of the two linear asymptotes shown (dashed lines) help to identify the two temporal regimes, before and after the transit time, t_{tr}.

composite before the transit time is not flat, but slopes downward gently. This is because some of the charges become immobilized temporarily within the material. These charges become mobile again at a later time, possibly after the transit time has elapsed. In this way the photocurrent after the transit time will fall away more slowly in a more dispersive medium. The area under the photocurrent transient represents the total quantity of charge photogenerated and released into the circuit. If losses due to long-term trapping are negligible, then total charge photogenerated is independent of the detail of the charge transport process. Accordingly if the slope of the photocurrent transient before t_{tr} is $-(1 - \alpha)$ where $0 \leq \alpha \leq 1$ is an empirical parameter of the material, then the slope of the transient after t_{tr} will be $-(1 + \alpha)$. As long as charges are only trapped temporarily, the two slopes will add to give -2, as they do for the photocurrent transient in a crystal. High dispersion corresponds to small values of α, whilst values of α close to unity correspond to a well-defined drift mobility for charge carriers.

Dispersive transport model of Scher and Montroll[1]

The main properties of dispersive transport described above can be deduced from a model of hopping transport between local sites each with a dwell time probability distribution

$$\psi(t) \propto t^{-(1+\alpha)} \; . \tag{5.3}$$

A simple approximation for modest electric field strengths is that the distance between localized hop sites occupied by a drifting carrier, $l(E) \propto E$ and in this case the mean position of a charge distribution in a TOF experiment

$$\bar{l}(t) \propto l(E)t^{\alpha} \; . \tag{5.4}$$

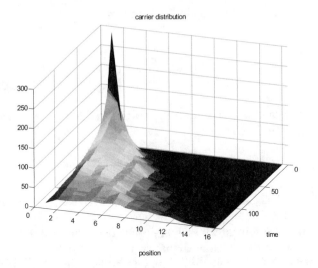

carrier distribution

position

time

Figure 5.5 Numerical Scher-Montroll calculation of the distribution of carriers over time, after their creation at a single location in a TOF experiment.

The peak of the charge distribution remains near the initial position, as shown in Figure 5.5. The result is that the variance of the charge distribution is proportional to the mean distance traveled. That is why the shape of the photocurrent transient is universal when normalized by transit parameters t_{tr} and i_{tr}; the long tail of the transient from dispersive media is not due to diffusion.

At high electric field strengths it has been argued to be better to allow a Boltzmann energy factor for forward and backward hops in the applied field, leading to the modification that the distance between hop sites for a drifting charge is now

$$l_{E high} \propto \sinh\left(\frac{e\rho E}{2kT}\right),$$
(5.5)

where ρ is the intersite distance for available hopping sites.[2]

The normal definition of mobility

$$\mu = \frac{d}{Et_{tr}}$$
(5.6)

can still be used for dispersive media, but now from (5.4) the ratio of mean position of charge to distance between hop sites

$$\frac{\bar{l}(E)}{l(E)} \propto t^{\alpha}$$
(5.7)

so the transit time is no longer linear with sample thickness, but instead

$$t_{tr} \propto \left[\frac{d}{l(E)} \right]^{1/\alpha}.$$ (5.8)

This leads to the calculation of mobility values in dispersive media that vary with field and sample size as

$$\mu \propto \left(\frac{d}{E} \right)^{1-1/\alpha}$$ (5.9)

dependent on the extent of the dispersion as described by the parameter α. When $\alpha = 1$ the limiting case of dispersion-free transport is obtained, but if α is closer to zero the extent of dispersion accounted for by the theory is higher. In brief, in the Scher-Montroll model for dispersive media such as amorphous photorefractive composites the mobility is not only a material parameter, it depends also on the material dispersion, α, and the experimental configuration, (d, E), used for the measurement.

Origins of dispersion in transport

In the Scher-Montroll model the dispersive transport properties arise when there is a balance between the probability of a delay for a carrier in transport and the consequence of that delay, such that the product delay × consequence is nearly constant. For example, the probability that a carrier is quickly trapped and remains so for the duration of a TOF experiment might be approximately the same as the probability that in the same time a carrier proceeds across the transport layer without any delays at all. Various intermediate outcomes are also equally likely in which the carrier makes some progress but also experiences some delays. The TOF experiment shows no characteristic time scale for the progress of carriers between the electrodes.

Dispersion-free transport may be observed in a TOF experiment if the sample is very thin. Here the probability of a delay is insufficient to lead to dispersion. Dispersion-free properties will also be observed if the range of delay times at hop sites is not wide enough to delay some carriers for a time comparable to the duration of the experiment.

The delays at hop sites can vary for several reasons. In amorphous media charge, transport is usually regarded as a hopping process between sites each of an individual energy. The absorption profiles of polymers and organic glasses are often broadened into Gaussian line shapes which are a signature of inhomogeneous broadening. For this reason, the distribution of transport site energy levels may also be considered to be Gaussian. Even in a medium with a homogeneous distribution of transport levels there may

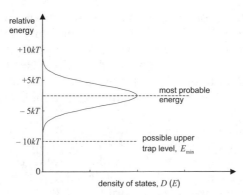

Figure 5.6 A Gaussian density of transport states describes the energetic distribution of levels subjected to inhomogeneous broadening. This could lead to an exponential distribution of low-lying trap states unlikely to be emptied thermally.

be an exponential distribution of defect or impurity levels beneath the transport manifold which may act as trapping sites.

If the low-energy tail of a Gaussian distribution is represented by an exponential density of states (Figure 5.6)

$$D(E) \propto \exp\left(\frac{E_{min} - E}{kT_0}\right) \tag{5.10}$$

where E_{min} is the lowest energy level for which delays are not significant, then a Boltzmann factor for the release time

$$t_{release} \propto \exp\left(\frac{E_{min} - E}{kT}\right) \tag{5.11}$$

will lead to dispersive transport due to the broad range of release times as long as the temperature $T << T_0$, the characteristic temperature of the density of states function. In such a situation the dispersion parameter $\alpha = T/T_0$.[3]

Gaussian disorder model

Monte Carlo simulations of the behavior of a system of carrier hopping sites subjected to energetic disorder, demonstrate that dispersion can explain temporal properties of transport in high applied electric fields.[4] An initial, high photocurrent falls to a plateau level that is nearly constant with time (as long as the width of the Gaussian density of states [DOS] distribution satisfies $\hat{\sigma} = \sigma/kT \leq 4.0$). This plateau is analogous to the steady current of a nondispersive transport medium. The current drops away again after the transit time. A shoulder is seen in the photocurrent transient as long as the dispersion is not too strong and this may be used to identify the transit time before the accelerated fall in the photocurrent which follows. Note that different behavior is predicted for a more dispersive system (such as at a cooler temperature), for which the width of the DOS $\sigma > 4kT$. Here the photocurrent falls continuously and there is no level plateau, although the rate of fall of the photocurrent does reduce on a double-logarithmic plot of current against time.

In the Gaussian disorder model we can physically identify three distinct regions corresponding to different stages in the evolution of the photocurrent transient. The initial high photocurrent corresponds to dominance of charge carriers with a high mobility, due to their photogeneration at energies higher than thermal equilibrium with the ambient would permit. These energetic carriers move through the transport site system at a level in the DOS that allows them many accessible transport sites. The second stage in the transient may be either level or a near-plateau of photocurrent falling gently over time. This corresponds to transport dominated by carriers at thermal equilibrium with the ambient as they move through the medium. The plateau or gentle slope ends at the transit time and this corresponds to the point at which perhaps 20% of the carriers have reached their destination electrode. After this time the photocurrent falls away as the number of carriers remaining in the transport region reduces.

In the Gaussian disorder model the individual energy levels of specific hopping sites in an amorphous medium *are assumed to be uncorrelated,* whereas hops to a lower energy are assumed not to be subject to any energetic barrier. Hops to sites of higher energy are assumed to follow the asymmetric Miller-Abrahams jump theory.[5] At thermal equilibrium a mobile carrier moving through a dipolar lattice of hopping sites regularly spaced with a Gaussian DOS is below the energetic center of the DOS by $-\sigma^2/kT$ where σ is the width of the DOS (Figure 5.7).[6] The activation energy needed

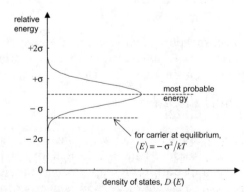

Figure 5.7 The thermal equilibrium energy level, $\langle E \rangle$, of a carrier within a Gaussian density of states is below the center of the distribution.

to release a carrier for conduction will decrease with temperature because the mean energy of the carriers increases with temperature. The mobility then has temperature dependence

$$\mu(T) = \mu_0 \exp\left\{-\left(\frac{2\sigma}{3kT}\right)^2\right\}. \tag{5.12}$$

Poole-Frenkel (root E) field dependence of mobility

There have been many experimental reports of the way mobility varies with electric field, E. In molecularly doped polymer under fields from 1–100 Vμm^{-1} the mobility μ is found to be proportional to $\exp\left(\gamma\sqrt{E}\right)$ where γ is a constant. A common elaboration is by Gill[7] that for PVK-based composites the mobility

$$\mu = \mu_0 \exp\left(\frac{-(E_0 - \beta\sqrt{E})}{kT_{eff}}\right) \tag{5.13}$$

is a function of an activation energy, E_0, for carrier conduction, the applied field, E, and an effective temperature given by

$$\frac{1}{T_{eff}} = \frac{1}{T} - \frac{1}{T_0}. \tag{5.14}$$

The often quoted Gill characteristic temperature, T_0, is found from an Arrhenius plot of mobility as a function of the reciprocal of temperature; lines obtained for different applied fields are found typically to converge at a reciprocal temperature $1/T_0$ as shown in Figure 5.8. The universal constant $\beta = 2.7 \times 10^{-5}$ eV(m/V)$^{1/2}$ over a wide range of experimental conditions.

Spatial (positional) disorder

Dispersive transport may arise even in a system with homogeneous absorption profiles if the distance between hopping sites varies sufficiently — known as spatial disorder. The tails of the wavefunctions for carriers at each end of the hopping sites may decay away with distance as $\Psi \propto \exp(-kz^2)$ such that the overlap between wavefunctions corresponding to adjacent sites (as in Figure 5.9) is a Gaussian distribution. If the distribution of intersite distances $\Delta z = z_2 - z_1$ is also Gaussian such that

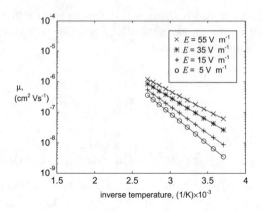

Figure 5.8 An Arrhenius plot shows the mobility, μ, on a logarithmic scale as a function of inverse temperature, $1/T$. Extrapolation of the lines obtained for different applied field strengths tends to indicate that a characteristic Gill temperature, T_0, exists for which mobility does not depend on electric field. Here $E_0 = 0.45$ eV, $T_0 = 500$ K.

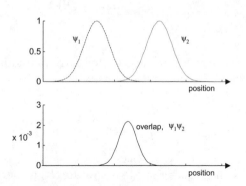

Figure 5.9 Configurational (positional) disorder leads to variations in the size of the overlap between wavefunctions of adjacent sites. The variation in hopping rate is in this case independent of temperature, unless a phase transition occurs such as due to heating above the glass transition temperature, T_g.

$$P(\Delta z) \propto \exp\left(\frac{-(\Delta z_0 - \Delta z)^2}{\sigma_{\Delta z}^2} \right) \tag{5.15}$$

then it is possible that the variation in hopping rate due to the variation in the overlap integral between adjacent sites is comparable to the corresponding variation in occurrence. Then dispersive transport will be observed. Note that in the absence of phase transitions, positional disorder would lead to a dispersion parameter independent of temperature while the dispersion due to energetic disorder is temperature dependent. As an example, PVK exhibits disorder that is temperature independent and thus may be argued to be mainly positional rather than energetic.

Spatial correlations in energy levels

According to convention, positional disorder has been introduced independently into the Gaussian disorder model by the parameter Σ, assumed to be uncorrelated with energetic disorder. For a medium in high applied field and using the Miller-Abrahams asymmetric jump theory, the mobility in a dispersive medium is predicted by the Gaussian disorder model to obey the relation

$$\mu(\hat{\sigma}, \Sigma, E) = \mu_0 \exp\left(-\frac{2}{3}\hat{\sigma}\right) \begin{cases} \exp C(\hat{\sigma}^2 - \Sigma^2)\sqrt{E} & \Sigma \geq 1.5 \\ \exp C(\hat{\sigma}^2 - 2.25)\sqrt{E} & \Sigma < 1.5 \end{cases} \qquad (5.16)$$

where $\hat{\sigma} = \dfrac{\sigma}{kT}$. The constant $C = 2.9 \times 10^{-4}$ (cm/V)$^{1/2}$.

As shown in the lower trace of Figure 5.10 the disorder model has had some success in reproducing the widely observed Poole-Frenkel dependence of mobility on $\sqrt{E}$.

However, it is now clear that spatial correlations do exist between neighbors in matrices containing dipoles such as molecularly doped polymers.[8] Relative energetic ridges and valleys are formed. Deeper valleys are wider, but less likely to be formed. One end result of this is that a Gaussian density

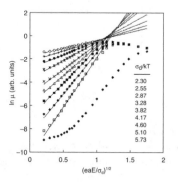

Figure 5.10 Field dependency of mobility in the correlated disorder model as a function of the width, σ_d/kT, of the density of states in the material. The lowest curve is the result of the Gaussian disorder model for $\sigma_d = 5.1kT$ where a limited region of linear Poole-Frenkel behavior is predicted. For typical values of site spacing, $a = 1nm$, for $\sigma_d = 100meV$ and a field strength of 100 Vμm^{-1} the parameter $eaE/\sigma_d \approx 1$ and the limit of the Poole-Frenkel behavior is reached. (With permission from S.V. Novikov, D.H. Dunlap, V.M. Kenkre, P.E. Parris and A.V. Vannikov, *Phys. Rev. Lett.*, 81(20) 4475–8 (1998). Copyright 1998 by the American Physical Society.)

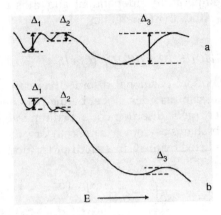

Figure 5.11 Correlated random potential with and without an applied field. Note that in the absence of a field the deepest potential wells are also the widest. When a field is applied in (b), wider potential wells are tilted more and hence experience much greater barrier reduction. (From Parris, P.E. et al., *Phys. Status Solidi B*, 218(1): 47–53, 2000. With permission.)

of states distribution is formed. Another result of the spatial correlation is that applied field assists in escape from deep and wide trap valleys much more than from shallow and narrow trap regions (Figure 5.11). Electric field reduces the dwell time in an energetic trap of radius r according to the drift Boltzmann factor $\exp(-eEr/kT)$. Where there is sufficient disorder the transport of charge is dominated by the escape rate from the traps with the longest dwell times. In strong poling fields the importance of the deepest and widest valleys is reduced and there is a radius characteristic of the traps with the longest dwell times,

$$r_c \propto \sqrt{\frac{\sigma^2}{2eEakT}} \tag{5.17}$$

that depends on the root of field.

A general expression for mobility at varying temperature and field strength is obtained from the correlated disorder model that

$$\mu = \mu_0 \exp\left[-\left(\frac{3\hat{\sigma}}{5}\right)^2 + C_0\left(\hat{\sigma}^{3/2} - \Gamma\right)\sqrt{\frac{eaE}{\sigma}}\right] \tag{5.18}$$

where for dipolar media $C_0 = 0.78$ and $\Gamma = 2$.

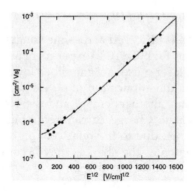

Figure 5.12 Measured field-dependent mobility for DEH-doped polycarbonate together with a theoretical fit to the data using the correlated disorder model. (Reprinted figure with permission from D.H. Dunlap, P.E. Parris and V.M. Kenkre, *Phys. Rev.
Lett.*, 77(3) 542–5 (1996). Copyright 1996 by the American Physical Society.)

How significant is dispersive transport for photorefraction?

How influential is the nature of the charge transport process on the photorefractive properties of amorphous organics? The standard model predicts
a minimum value of the mobility, above which the rate of holographic
response is determined by charge generation rather than transport. It seems
likely that the higher mobility values of some amorphous composites (those
with mobility $\mu \geq 10^{-6}$ cm^2/Vs) should be sufficient for the charge generation
limit to be approached closely in most typical configurations.[9] The standard
model was developed for the case of a single-valued mobility, however. The
dispersive transport process corresponds to apparent mobility values that
may depend on the size of the transport region and the moment in time to
be considered. Short propagation distances or times may yield higher apparent mobilities. Over time, the mobility will appear to fall during a dispersive
transport process as the number of hot carriers declines. Clearly specific
conditions must be considered before mobility can be disregarded as a significant dynamic factor in photorefraction in amorphous organics.

In the steady state of the saturated photorefractive hologram, the field
dependence of the mobility might be expected to influence either the contrast
or the phase of the space-charge field pattern formed by an intensity pattern.
Schildkraut and coworkers both in numerical and analytical work have
modeled the effect of the field dependence of the mobility.[10] Using typical
parameters for mobility field dependence led to only very slight modifications to the space-charge field profile.

Holographic time-of-flight experiments

The transport of charge carriers is observed across the fringes of an optical pattern in an holographic time-of-flight (HTOF) experiment as in Figure 5.13. Two picosecond or nanosecond laser pulses are combined interferometrically within a photoconductor under the application of a field. In the regions of high intensity charge carriers are photogenerated and these drift in a field toward the adjacent bright fringe. Charge carrier recombination may be dominant at the next bright fringe due to the relatively high concentration of sensitizer ions there. Alternatively, the charges may become distributed fairly evenly through the medium due to dispersive transport, if carrier recombination is not dominant.

If a medium with a high glass transition temperature, T_g, is used, then the Pockels electro-optic effect will be the dominant cause of refractive index modulation in the hologram. The Pockels effect is very fast, and the evolution of diffraction from the hologram over time will indicate the evolution of the space-charge field pattern.

A single-diffraction peak was observed in one HTOF experiment[11] with such a high T_g composite (Figure 5.14). In this example it seems likely that some charge carriers were able to drift more than 5 μm through the medium prior to recombination. Thus they moved from one bright fringe past a position of anti-coincidence with the bright fringes. The space-charge field is maximized at anti-coincidence. The carriers continued to drift until their dispersion led to a near-uniform distribution of carriers, leading to a reduced but steady space-charge field that persisted due to the patterning of the sensitizer ion population.

It is desirable to use HTOF to observe mobility over holographic length scales in low T_g composites, but the relationship between a dynamic space-charge field and the birefringence due to reorientational diffusion is complicated. Rotational effects are capable of masking any diffraction maxima due to short-lived peaks in the space-charge field amplitude that might otherwise have occurred. In order to observe the evolution of the

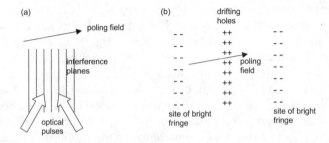

Figure 5.13 The holographic time-of-flight (HTOF) experiment. (a) Two fast optical pulses interfere to produce a short-lived interference pattern. (b) Charges drift across the pattern, producing an electric field pattern which leads to optical diffraction of a probe beam.

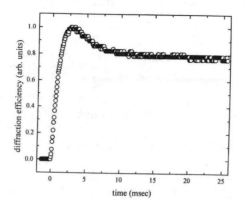

Figure 5.14 Experimental HTOF diffraction efficiency over time, for a photorefractive composite with a high T_g, in which reorientation of chromophores in response to the optical stimulus was negligible. (Reprinted figure with permission from G.G. Malliaras, V.V. Krasnikov, H.J. Bolink and G. Hadziioannou, *Phys. Rev. B*, 52(20) R14 324–7, (1995). Copyright 1995 by the American Physical Society.)

space-charge field reliably it is necessary that the reorientation process be either much faster or slower than the evolution of the space-charge field. The latter condition corresponds to high T_g materials. The former condition, that reorientation is clearly faster, may be difficult to satisfy in amorphous composites because both processes can be dispersive with a broad range of time scales. Experimentally, it is important to ensure that conductivity and the erasure of the space-charge field pattern by the probe beam do not create a peak of holographic diffraction efficiency that may be mistaken for the diffraction peak due to carrier anti-coincidence.

Langevin trapping and photorefractive dynamics almost independent of drift mobility

Photorefractive composites are usually doped with a sensitizer to ensure photoconductive response in the visible spectrum. The sensitizer sites are photoionized to a density normally exceeding 10^{16} cm^{-3}. The effect of charge carrier drift mobility may be reduced greatly for the case of a photorefractive medium in which the carrier trapping process is predominately at charged sites.

Consider the usual case that quantum efficiency of charge photogeneration is small relative to unity. The dominant effect of mobility on the dynamics of the photorefraction process described in chapter 3 is through the effect of geminate recombination due to spatially averaged hole and electron mobility, $\langle \mu \rangle$, on photogeneration rate, s, as

$$s \propto k_d(E) \propto \frac{\langle \mu \rangle e}{\varepsilon_s} . \qquad (5.19)$$

That this is so can be shown from the expression in chapter 4 for the space-charge field rise time

$$\tau_{rise} = \frac{\varepsilon_s}{p_0 e \mu} \frac{(E_M + jE_0)}{E_M} \frac{E_q}{([1 + p\cos^2 \phi]E_q + jE_0)} \tag{5.20}$$

in which several terms may be simplified in the case of Langevin (charged) trap sites.

First recall that the saturated hole density is

$$p_0 = sI_0(N_A - N_D^+)\tau_h \tag{5.21}$$

The mobility field is the field necessary for a carrier to travel one radian through a sinusoidal grating in one carrier lifetime,

$$E_M = (\mu\tau_h K)^{-1} , \tag{5.22}$$

and depends on the drift mobility, μ, of the mobile carrier (hole). Substituting for the mean hole lifetime $\tau_h = (\gamma_R N_D^+)^{-1}$ we obtain

$$E_M = \frac{\gamma_R N_D^+}{\mu K} . \tag{5.23}$$

The carrier recombination coefficient for trapping at singly charged sites

$$\gamma_R = \frac{\langle\mu\rangle e}{\varepsilon_s} \tag{5.24}$$

depends on the spatial average, $\langle\mu\rangle$, of electron and hole mobilities. In the single-carrier model of photorefraction the holes are regarded as mobile while the negative charges are held immobile at sensitizer sites. For this reason the spatial average of mobilities is likely to be dominated by the hole mobility and to a close approximation the ratio $\langle\mu\rangle/\mu$ is of order unity and independent of the value of the hole mobility itself.

We can conclude that in the Langevin trapping case the mobility field

$$E_M\big|_{Langevin} = \frac{eN_D^+}{\varepsilon_s K}\frac{\langle\mu\rangle}{\mu} \tag{5.25}$$

is in fact almost independent of the hole mobility. Physically this is because the effect of improved hole drift mobility is to increase the rate of hole recombination with immobile traps such that the product of mobility and lifetime is approximately independent of the mobility itself. A greater mobility will increase the average speed of drifting carriers but also shortens the carrier lifetime prior to recombination at a charged trap site. Recalling that the general expression for the saturation field is

$$E_q = \frac{eN_D^+}{\varepsilon_s K}\left(1 - \frac{N_D^+}{N_A}\right) \tag{5.26}$$

we can note that in the special case of Langevin trapping the simple relation

$$E_q\big|_{Langevin} = \left(1 - \frac{N_D^+}{N_A}\right)\frac{\mu}{\langle\mu\rangle}E_M\big|_{Langevin} \tag{5.27}$$

will hold. These substitutions yield for the space-charge field rise time in the Langevin case

$$\tau_{rise}\big|_{Langevin} = \frac{N_D^+}{sI_0(N_A - N_D^+)}\frac{\langle\mu\rangle}{\mu}\left[1 + \frac{jE_0}{E_M\big|_{Langevin}}\right]\frac{E_q\big|_{Langevin}}{([1 + p\cos^2\phi]E_q\big|_{Langevin} + jE_0)} \tag{5.28}$$

which rearranges in terms of the saturation field to give

$$\tau_{rise}\big|_{Langevin} = \frac{N_D^+}{sI_0 N_A}\left\{\frac{\dfrac{\langle\mu\rangle}{\mu}\dfrac{N_A}{(N_A - N_D^+)} + j\left(\dfrac{E_0}{E_q\big|_{Langevin}}\right)}{1 + p\cos^2\phi + j\left(\dfrac{E_0}{E_q\big|_{Langevin}}\right)}\right\}. \tag{5.29}$$

So in the case of exclusively Langevin (singly charged) trap density, the rise time of the space-charge field is dependent on the initial concentration of the compensators, N_D^+, the applied field strength, E_0 and the angle ϕ between the grating vector and poling field. This is along with the more obvious dependence on grating vector, K (through the mobility field), and the photogeneration rate factor sI_0.

Where Langevin trapping is the dominant trapping process within organic photorefractive materials the study of efficiency of photogeneration of charge is most relevant to the improvement of the rise time for hologram formation. While mobility will influence the photogeneration process, it may be that the effect of any improvement in drift mobility on the remainder of the dynamics of the photorefractive process will only be through any consequent change in the ratio between hole mobility and the spatial average mobility that governs recombination.

Drift mobility dependence will be observed in the photorefractive charge transport process if there are significant numbers of uncharged trap sites. The carrier lifetime, in this case, would depend on the cumulative cross section of neutral traps (dependent on number density as $N^{2/3}$) rather than the trap density and the influence of mobility would not cancel out so completely.

Example

In a simple case, the applied field component along the grating vector, E_0, is smaller than either the mobility or saturation fields, $E_0 << E_M, E_q$. Neglecting terms in applied field is equivalent to using vanishingly small grating vector, K. Together with the usual assumption that the acceptor sites are only slightly ionized, $N_D^+ << N_A$ the Langevin expresson for the rise time can be approximated quite well as

$$\tau_{rise}\big|_{Langevin} \approx \frac{N_D^+}{sI_0(1+p\cos^2\phi)(N_A-N_D^+)}\frac{\langle\mu\rangle}{\mu} \tag{5.30}$$

or very approximately for a common arrangement

$$\tau_{rise}\big|_{Langevin} \approx\approx \frac{1}{sI_0}\frac{N_D^+}{2N_A}\frac{\langle\mu\rangle}{\mu}. \tag{5.31}$$

From this, it is clear that a fast response will be obtained for a material with high photosensitivity (high s) and very poor sensitizer ion mobility such that $\langle\mu\rangle/\mu$ is minimized but the mobility itself is only required to be sufficient that $E_0 << E_M, E_q$.

Charge generation limit to buildup of space-charge field

The charge generation limit to the rate of rise of the space-charge field was given in chapter 4 as

$$\frac{\partial}{\partial t}\Delta E \le e\frac{\phi}{\hbar\omega}\frac{\alpha mI_0}{\varepsilon_s K}\left(\frac{-E_0}{E_M+jE_0}\right) \tag{5.32}$$

and with the Langevin value of the mobility field given above, it is clear that the fastest buildup of space-charge field will be obtained usually in the case of low-trap density and grating period (i.e., minimized value of N_D^+/K).

References

1. Scher, H. and Montroll, E.W., Anomalous transit-time dispersion in amorphous solids, *Phys. Rev. B.*, 12(6): 2455–2477, 1975.
2. Pfister, G., Dispersive (non-Gaussian) transient transport in disordered solids, *Adv. Phys.*, 27: 747, 1978.
3. Scher, H., Shlesinger, M.F., and Bendler, J.T., Time-scale invariance in transport and relaxation, *Phys. Today,* 44(1): 26–34, 1991.
4. Bässler, H., Charge transport in disordered organic photoconductors — a Monte Carlo simulation study, *Phys. Status Solidi B*, 175: 15; 1993; Charge transport in molecularity doped polymers, *Philos. Mag.*, 50: 347, 1984.
5. Miller, A. and Abrahams, E., Impurity conduction at low concentrations, *Phys. Rev.*, 120: 745, 1960.
6. Dieckmann, A., Bässler, H., and Borsenberger, P.M., An assessment of the role of dipoles on the density-of-states function of disordered molecular solids, *J. Chem. Phys.*, 99(10): 8136–8141, 1993.
7. Gill, W.D., Drift mobilities in amorphous charge transfer complexes of trinitrofluorenone and poly N-vinylcarbazole, *J. Appl. Phys.*, 43(12) 5033–5040 (1972).
8. Parris, P.E., Dunlap, D.H., and Kenkre, V.M., Energetic disorder, spatial correlations, and the high field mobility of injected charge carriers in organic solids, *Phys. Status Solidi B*, 218(1): 47–53, 2000.
9. West, D.P., Rahn, M.D., Im, C., and Bässler, H., Hole transport through chromophores in a photorefractive polymer composite based on poly(N-vinylcarbazole), *Chem. Phys. Lett.*, 326(5–6): 407–412, 2000.
10. Schildkraut, J.S. and Buettner, A.V., Theory and simulation of the formation and erasure of space-charge gratings in photoconductive polymers, *J. Appl. Phys.*, 72(5): 1888–1893, 1992; Schildkraut, J.S. and Cui, Y., Zero-order and first-order theory of the formation of space-charge gratings in photoconductive polymers, *J. Appl. Phys.*, 72(11): 5055–5060, 1992.
11. Malliaras, G.G., Krasnikov, V.V., Bolink, H.J., and Hadziioannou, G., Holographic time of flight measurements of the hole drift mobility in a photorefractive polymer, *Phys. Rev. B*, 52(20): R14, 324–327, 1995.

chapter six

Steady-state electro-optics in amorphous photorefractive composites with reorientational effects

Contrast in the refractive index is the dominant cause of diffraction of light from holograms in photorefractive materials. In photorefractive polymer composites this refractive index contrast depends on the electric space-charge field pattern. This electric field pattern controls the refractive index through the electro-optic effect.

Theory of electro-optic response

Light propagates according to the wave equation for the optical field, $\underline{E}$, that

$$\nabla^2 \underline{E} - \left(\varepsilon/c^2\right)\frac{\partial^2}{\partial t^2}\underline{E} = 0 \tag{6.1}$$

where the dielectric constant, ε , is used to describe the reduction of the phase velocity in a polarizable medium, away from the vacuum value, c. The polarization vector $\underline{P}$ may not be parallel to the optical field vector, $\underline{E}$, in an anisotropic medium. For this reason $\underline{\varepsilon}$ is a 3×3 tensor, related to the susceptibility by $\underline{\chi} = \underline{\varepsilon} - 1$, that describes the polarization according to

$$\underline{P} = \varepsilon_0 \underline{\chi} \cdot \underline{E} . \tag{6.2}$$

The refractive index, n, is the scalar ratio by which a wave (with a specific orientation of the optical field vector) has been slowed down. For that specific wave the scalar relation

$$n^2 = \chi_{eff} + 1 \tag{6.3}$$

holds between the index and the effective susceptibility for that wave. Modulations of the refractive index, Δn, may then be calculated according to

$$\Delta n = \frac{\Delta \chi_{eff}}{2n} \tag{6.4}$$

where the modulation of the electric field has led to a modulation of the susceptibility of $\Delta \chi_{eff}$.

In three dimensions there are three orthogonal components of the optical field vector. If these are selected to diagonalize the dielectric tensor, $\overset{=}{\varepsilon}$, then three principal axes for the material are identified. An optical field with an arbitrary orientation may be projected onto these three principal axes to calculate the polarization component along each of them. These polarization components may be in turn projected onto the field vector of the emergent wave to obtain the properties of that wave, as in Figure 6.1.

Where incident and emergent light are collinear it follows that for light with electric vector projected at angles ϑ_x, ϑ_y, ϑ_z from the three principal axes of the material the effective dielectric constant $\varepsilon(\vartheta_x, \vartheta_y, \vartheta_z)$ will be obtained from the summation[1]

$$\frac{1}{\varepsilon\left(\vartheta_x, \vartheta_y, \vartheta_z\right)} = \sum_{i=x,y,z} \frac{\cos^2 \vartheta_i}{\varepsilon_{ii}} \tag{6.5}$$

which defines an ellipsoid.

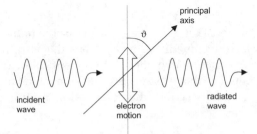

Figure 6.1 Propagation of a light wave in matter. The oscillating electric field of the incoming wave drives predominantly electronic motion that then radiates a scattered wave. Interference of the scattered wave with the transmitted incident wave leads to a phase delay as if light traveled more slowly.

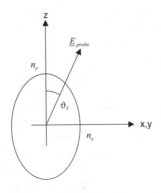

Figure 6.2 The index ellipse is a radial plot of the refractive index in a uniaxial material as a function of angle, ϑ_2, between the one, unique principal axis and the electric vector of the probe wave field.

There is cylindrical symmetry in an amorphous medium with a poling field. Two of the principal axes are indistinguishable in this (uniaxial) case, the projections onto the x and y axes become unimportant with the result that

$$\frac{1}{n^2(\vartheta_z)} = \frac{\cos^2 \vartheta_z}{n_e^2} + \frac{\sin^2 \vartheta_z}{n_o^2}. \qquad (6.6)$$

Equation (6.6) defines an ellipse for the refractive index, $n(\vartheta_z)$, experienced by the wave dependent on the orientation of the optical field relative to the poling field directed along the z axis as shown in Figure 6.2. The notation $n_e^2 = \varepsilon_{zz}$ and $n_o^2 = \varepsilon_{xx} = \varepsilon_{yy}$ has been used. A useful first-order approximation of the ellipse equation is

$$n(\vartheta_z) \approx (n_e - n_o)\cos^2 \vartheta_z + n_o. \qquad (6.7)$$

Although it is more accurate to retain the proper index ellipsoid equation when calculating the refractive index within a photorefractive hologram, the error introduced by the simplification to first order in (6.7) can be shown to be negligible in the likely case that the birefringence $n_e - n_o \le 10^{-3}$. The electro-optic (EO) effect leads to a change in this refractive index due to changes in the applied poling field. In the nonresonant case of weak absorption a Taylor series expansion is useful to describe the change in the dielectric constant due to the field

$$\Delta\left(\frac{1}{\varepsilon}\right)_{E_{pol}} = rE_{pol} + sE_{pol}^2 + ... \qquad (6.8)$$

where a scalar form has been used for clarity. The linear term on the right-hand side is the Pockels, linear EO effect and the quadratic term is due to Kerr response. The reversal of the applied field reverses the sense in which refraction is altered for the linear Pockels term, so this term is zero in centrosymmetric media.

Amorphous photorefractive materials contain dipolar molecules or moieties that are oriented by an applied field. Individual dipoles are anisotropic in their polarizability, which is generally greater for fields oriented parallel to the long axis of the molecule and reduced for fields across the width of the dipolar unit. In addition to this, their dipolar nature is associated with the non-centrosymmetry that is needed for the Pockels linear EO effect. For these two distinct reasons, a change in the poling field will lead to a change in refraction and also a changed distribution of orientations if the dipoles have the freedom to rotate in response to the field.

Using the assumption that each dipolar molecule may be represented by a vector electric dipole, μ , with dipole energy $-\mu \cdot \underline{E}$ in the applied field, competition between orientational potential energy and thermal energy will lead to a Maxwell-Boltzmann distribution of dipolar orientations to the field. The balance parameter, $u = \mu E / kT$ is the normalization of the dipole energy by the thermal energy. In terms of this, we can express the properties of the orientational distribution of dipoles at thermal equilibrium in terms of the spherical modified Bessel functions, $i_n(u)$:[2]

$$i_0(u) = \frac{\sinh u}{u}$$

$$i_1(u) = \frac{u \cosh u - \sinh u}{u^2} \quad . \tag{6.9}$$

$$i_n(u) = \frac{u}{2n+1} \left[i_{n-1}(u) - i_{n+1}(u) \right]$$

The order parameters, A_n, used to describe the orientational ordering of the medium are derived from these by the relation

$$A_n(u) = \frac{i_n(u)}{i_0(u)} \tag{6.10}$$

so that the first three parameters are

$$A_0 = 1$$

$$A_1 = \frac{u \cosh u - \sinh u}{u \sinh u} \approx \frac{5u}{15 + u^2} \tag{6.11}$$

$$A_2 = 1 - \frac{3}{u}\left(\frac{u \cosh u - \sinh u}{u \sinh u}\right) \approx \frac{u^2}{15 + u^2}$$

where the quadratic approximations given are useful when $u < 1$.

These order parameters are used to calculate the polarizability of the distribution of N-oriented dipoles according to the relations

$$\chi_{zz} = N\bar{\alpha}\left(1 + 2\gamma A_2\right)$$
$$\chi_{xx} = N\bar{\alpha}\left(1 - \gamma A_2\right) \tag{6.12}$$

in which

$$\gamma = \frac{\alpha_{//} - \alpha_\perp}{\alpha_{//} + 2\alpha_\perp} \tag{6.13}$$

is the anisotropy in the molecular polarizability, and

$$\bar{\alpha} = \frac{\alpha_{//} + 2\alpha_\perp}{3} \tag{6.14}$$

is the orientational average of the molecular polarizability. The susceptibility, χ, in equation 6.12 is a function of the second-order parameter, A_2, because if the dipoles are free to rotate then the orientational distribution of the dipoles depends on the poling field. The Pockels linear EO response is described by

$$r_{zzz} = N\beta\left(\frac{3}{5}A_1 + \frac{2}{5}A_3\right) \approx N\beta\frac{3}{5}A_1$$
$$r_{xxz} = N\beta\left(\frac{1}{5}A_1 - \frac{1}{5}A_3\right) \approx N\beta\frac{1}{5}A_1 \tag{6.15}$$

which depends on the first molecular hyperpolarizability of the dipole, β. Under normal circumstances the third-order parameter $A_3 \approx 0$ (derived from equations 6.9 and 6.10) for fields weaker than the dielectric breakdown field for the photorefractive composites. The Pockels EO response is thus dependent primarily on the first-order parameter, A_1.

Equation (6.12) describes the effect on the susceptibility of the reorientation of dipolar molecules on the polarizability due to their anisotropy alone, often referred to as the reorientational effect. Equations (6.15) and (6.8) together describe the Pockels nonlinearity, sometimes called the simple EO effect in the context of amorphous photorefractive materials. The two effects both tend to increase the refractive index for light polarized along the poling field. Perpendicular to this axis the reorientational effect and the Pockels effect act in opposite senses. Increased poling field leads to a reduction in χ_{xx} and a corresponding fall in refractive index, while typically r_{xxz} is increased leading to a corresponding rise in index, as illustrated in Figure 6.3.

The orientation of dipoles will be fastest when an amorphous material is above the glass transition temperature, T_g. If the temperature of the material is allowed to cool to below T_g after poling, then significant further reorientation will not occur. Pockels electro-optic response to the space-charge field will then dominate the photorefractive effect with a nonlinear coefficient determined by the poling field. In this case the electro-optic change in index will depend on the product of the first-order parameter and the field, so that the index change

$$\Delta n_{Pockels} \propto \frac{5u}{15 + u^2} \cdot u \, , \tag{6.16}$$

is approximately quadratic with field for $u < 1$. In fact the change in refractive index at angle ϑ_2 will be given to first order by

$$\Delta n \left(\vartheta_z, u \right)_{Pockels} \approx \left(\frac{1}{2n} \right) \frac{2u^2}{15} \left(\cos^2 \vartheta_z + \frac{1}{2} \right) N\beta \left(\frac{kT}{\mu} \right) . \tag{6.17}$$

If a composite or glass is used at a temperature $T \geq T_g$ so that dipolar reorientation occurs during the photorefractive process, then the electro-optic change in index due to this reorientation of anisotropic dipoles will be dependent on the second-order parameter so that the index change

$$\Delta n_{reorient} \propto \frac{u^2}{15 + u^2} \tag{6.18}$$

for $u < 1$. The refractive index change due to this is given to first order by

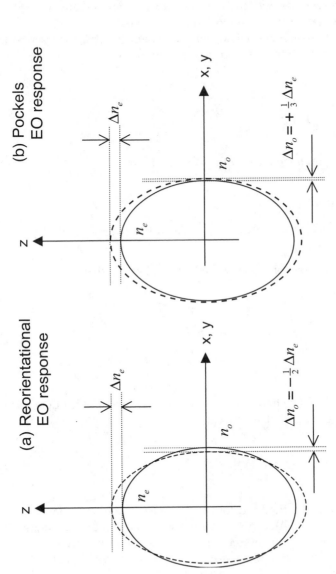

Figure 6.3 A comparison of reorientational and Pockels ellipsoid changes as an electric field is increased. The volume of the ellipsoid is a constant in the case of the reorientational effect (a), in which the polarizability of the chromophores is redistributed orientationally. The ellipsoid changes size in a Pockels change to the polarizability (b), as the polarizability may be enhanced due to the effect of the electric field on the electron populations of the chromophores.

$$\Delta n\left(\vartheta_z,u\right)_{T \geq T_g} \approx \left(\frac{1}{2n}\right)\frac{u^2}{15}\left(\cos^2\vartheta_z - \frac{1}{3}\right)N\left(\alpha_{//}-\alpha_{\perp}\right). \tag{6.19}$$

This term dominates over the Pockels contribution in many amorphous photorefractive materials where reorientation occurs.

It has become conventional to define a coefficient, C, to describe the change in index due to the applied field in the case of $\vartheta_2 = 0$, where

$$C=\left[\frac{1}{15}\left(\frac{2}{3}\right)N\left(\alpha_{//}-\alpha_{\perp}\right)+\frac{2}{15}\left(\frac{3}{2}\right)N\beta\frac{kT}{\mu}\right]\left(\frac{\mu}{kT}\right)^2 \tag{6.20}$$

and a coefficient, A, for the case of $\vartheta_2 = 90°$, where

$$A=\left[\frac{1}{15}\left(-\frac{1}{3}\right)N\left(\alpha_{//}-\alpha_{\perp}\right)+\frac{2}{15}\left(\frac{1}{2}\right)N\beta\frac{kT}{\mu}\right]\left(\frac{\mu}{kT}\right)^2. \tag{6.21}$$

The index changes for these two limiting cases are then given by

$$\Delta n_e\left(E\right)=\frac{1}{2n}CE^2$$
$$\Delta n_o\left(E\right)=\frac{1}{2n}AE^2 \tag{6.22}$$

and in this notation the change in the ordinary refractive index depends on the total field-induced birefringence as

$$\Delta n_o\left(E\right)=\frac{A}{C-A}\left(n_e-n_o\right). \tag{6.23}$$

The fraction of the field-induced birefringence due to a change in the ordinary index, $A/(C-A)$, can vary between $-1/3$ and $+1/2$, dependent on the relative significance of the reorientational anisotropic index variations and the Pockels EO effect in the material. A generalized version of (6.17) and (6.19) can be written using this notation, that

$$\Delta n\left(\vartheta_z,u\right)_{T \geq T_g} \approx \frac{1}{2n}\left(C-A\right)E^2\left(\cos^2\vartheta_z + \frac{A}{C-A}\right) \tag{6.24}$$

Holographic diffraction efficiency

Light propagating through a transparent, polarizable medium has a reduced phase velocity, c/n, which is the result of the optical frequency polarization of the medium. Calculation of the polarization involves the projection of the optical field onto the principal axes of the medium (i.e., the dye molecules) and the projection of this polarization back onto the optical field vector of the radiated wave.

The polarizability of the medium is patterned sinusoidally by a space-charge field in a photorefractive medium. The pattern leads to Bragg diffraction of light into an emergent wave with a new direction determined by the Bragg relation for the scatter angle, $\Delta\beta$, that

$$\lambda = 2\Lambda \sin \Delta\beta \tag{6.25}$$

where Λ is the period of the pattern and λ is the optical wavelength. In order to calculate the diffraction efficiency, η, into the scattered wave it is necessary to calculate the projections of the optical incident wave, $\underline{E}_{in}$, onto the polarizability pattern, $\Delta\underline{\chi}$, and of that onto the field vector of the scattered wave, $\underline{E}_{out}$. Although care should be taken to ensure its applicability, commonly the diffraction efficiency is calculated with acceptable accuracy for photorefractive polymers according to the simple Kogelnik formula

$$\eta = \sin^2\left(\frac{\pi\Delta nL}{\lambda} \left(\underline{E}_{out}^* \cdot \underline{E}_{in} \right) \right) \tag{6.26}$$

where the interaction length, L, depends on the sample thickness, d, as

$$L = \frac{d}{\sqrt{\cos\beta \cos\left(\beta + \Delta\beta\right)}} . \tag{6.27}$$

The dot product $\underline{E}_{out}^* \cdot \underline{E}_{in}$ is needed to account for the non-unity projection of the polarization unit vector of the wave incident onto the field vector of the wave scattered, if the two waves interfering do not have parallel electric field vectors. This is possible with p-polarized light, in which the electric vectors are in the same plane of incidence and diffraction. All electric vectors will be aligned parallel in the case of s-polarization (where all oscillating electric fields are polarized perpendicular to the plane of incidence and diffraction).

Formally, the index contrast

$$\Delta n = \frac{1}{2n}\left(\underline{E}_{out}^* \cdot \Delta\underline{\chi} \cdot \underline{E}_{in} \right) \tag{6.28}$$

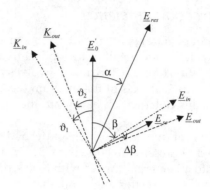

Figure 6.4 Geometry of optical vectors and fields in holographic four wave mixing, using the notation of the text.

and in the simple case of s-polarization, a scalar form of this will be sufficient. The optical fields will all be perpendicular to the space-charge field, which will be in the plane of incidence and diffraction. If the poling field is also oriented in this plane then the susceptibility pattern will depend on the resultant of the space-charge and poling fields as

$$\Delta\chi_{s-pol} = A E_{res}^2 \tag{6.29}$$

and identifying the first spatial Fourier (1K) component of the pattern to be, as in Figure 6.4,

$$\Delta\chi_{1K}\big|_{s-pol} = A \cdot 2 E_0' E_{sc} \cos\phi \ , \tag{6.30}$$

where ϕ is the angle subtended between the two fields. The 1K component of the index contrast observed for s-polarized optical diffraction will be given by

$$\Delta n_{1K}\big|_{s-pol} = \frac{1}{2n} \cdot 2 A E_0' E_{sc} \cos\phi \ . \tag{6.31}$$

In the case of p-polarization of the beams, the electric vectors of the different waves are no longer parallel. In the local frame in which $\underline{E}_{res}$ is along z, the susceptibility modulation should be

$$\underline{\underline{\Delta\chi}}_{local} = \begin{bmatrix} A & 0 & 0 \\ 0 & A & 0 \\ 0 & 0 & C \end{bmatrix} E_{res}^2 \tag{6.32}$$

but the angle, α, between the poling field and the resultant field varies with the space-charge field strength through the hologram. Using the rotation matrix

$$\underline{R} = \begin{bmatrix} \cos\alpha & 0 & -\sin\alpha \\ 0 & 1 & 0 \\ \sin\alpha & 0 & \cos\alpha \end{bmatrix} \tag{6.33}$$

the susceptibility modulation can be reexpressed in the device frame, in which the poling field is along z, as

$$\underline{\Delta\chi}_{device} = \underline{R}^T \cdot \underline{\Delta\chi}_{local} \cdot \underline{R} \tag{6.34}$$

which has the 1K component

$$\underline{\Delta\chi}_{1K}\Big|_{p-pol} = \begin{bmatrix} 2A\cos\phi & 0 & (C-A)\sin\phi \\ 0 & 2A\cos\phi & 0 \\ (C-A)\sin\phi & 0 & 2C\cos\phi \end{bmatrix} E'_0 E_{sc} . \tag{6.35}$$

While the second row of this matrix (y-coordinate) reconfirms the case for s-polarized light, in the p-polarized case (x,z) the index contrast is predicted to be

$$\Delta n_{1K}\Big|_{p-pol} = \frac{1}{2n} \begin{pmatrix} \cos\vartheta_2 & 0 & \sin\vartheta_2 \end{pmatrix}$$
$$\begin{bmatrix} 2A\cos\phi & 0 & (C-A)\sin\phi \\ 0 & 2A\cos\phi & 0 \\ (C-A)\sin\phi & 0 & 2C\cos\phi \end{bmatrix} \begin{pmatrix} \cos\vartheta_1 \\ 0 \\ \sin\vartheta_1 \end{pmatrix} E'_0 E_{sc} \tag{6.36}$$

or alternatively

$$\Delta n_{1K}\Big|_{p-pol} = \frac{1}{2n} \Big(2A\cos\phi\cos\vartheta_1\cos\vartheta_2 + (C-A)\sin\phi\sin(\vartheta_1+\vartheta_2) + \tag{6.37}$$
$$2C\cos\phi\sin\vartheta_1\sin\vartheta_2 \Big) E'_0 E_{sc}$$

The probe wave frame for holographic diffraction

It is useful in the study of amorphous media to transform the description of diffraction of light with p-polarization into the frame of reference of the probe optical wave as it propagates through the hologram. For a probe with optical field vector at an angle, β, to the poling field, the projection of the field onto the direction of $\underline{E}_{res}$ is $\cos(\alpha - \beta)$ in the local frame. This leads to a radiated contribution to the emergent wave of $C \cos(\alpha - \beta) \cos(2\phi - \beta - \alpha)$. The contribution corresponding to the projection of the probe field onto the perpendicular axis of the local frame is $A(-\sin(\alpha - \beta)) \sin(2\phi - \beta - \alpha)$. In terms of the scatter angle

$$\Delta\beta = 2(\phi - \beta) \tag{6.38}$$

the effective index contrast for p-polarized light is therefore

$$\Delta n_{p-pol} = \frac{1}{2n}$$
$$\left[C \cos(\beta - \alpha) \cos(\beta + \Delta\beta - \alpha) + A \sin(\beta - \alpha) \sin(\beta + \Delta\beta - \alpha) \right] E_{res}^2 \tag{6.39}$$

Using the double-angle formulae

$$\cos(\Delta\beta + \beta - \alpha) = \cos\Delta\beta \cos(\beta - \alpha) - \sin\Delta\beta \sin(\beta - \alpha)$$
$$\sin(\Delta\beta + \beta - \alpha) = \sin\Delta\beta \cos(\beta - \alpha) + \cos\Delta\beta \sin(\beta - \alpha) \tag{6.40}$$

this can be rearranged to

$$\Delta n_{p-pol} = \frac{1}{2n}(C - A)E_{res}^2 \cdot$$
$$\cos\Delta\beta \left[\frac{A}{C - A} + \cos^2(\beta - \alpha) - \tan\Delta\beta \frac{\sin 2(\beta - \alpha)}{2} \right] \cdot \tag{6.41}$$

The first two terms in the square bracket may be recognized as the index contrast experienced by the probe wave as it propagates straight through the medium, undiffracted. These two terms represent here the contribution to the diffracted wave due to the transverse material polarization induced by the probe wave. The third term is a relatively small contribution to the diffracted wave from the longitudinal component of polarization induced by the probe wave.

Noting that

$$\frac{1}{2}\sin 2(\beta - \alpha) = \cos(\beta - \alpha)\sin(\beta - \alpha) \tag{6.42}$$

and that

$$\sin \alpha = \frac{E_{sc} \sin \phi \sin Kx}{E_{res}} \tag{6.43}$$

the 1K component of $E_{res}^2 \left(\frac{1}{2}\right)\sin 2(\beta - \alpha)$ is

$$2E_0'E_{sc} \cos \phi \cos \beta \sin \beta - E_0'E_{sc} \sin \phi \cos 2\beta \tag{6.44}$$

and the first spatial Fourier component of the index contrast for diffraction of p-polarized light

$$\Delta n_{1K}\Big|_{p-pol} = \frac{1}{2n}(C - A) \cdot 2E_0'E_{sc} \cos \Delta\beta \left(\begin{array}{c} \dfrac{A}{C - A}\cos \phi + \cos(\phi - \beta)\cos \beta \\[2mm] + \dfrac{\tan \Delta\beta}{2}\sin(2\beta - \phi) \end{array} \right), \tag{6.45}$$

an expression which is equivalent to the expression in equation 6.36, which was obtained using the device (poling field) frame of reference. The two expressions vary slightly at high levels of birefringence due to the use of a first-order expression for the index ellipsoid in the development of (6.45), but this difference is not significant where holographic index contrast is below 5×10^{-3}.[3]

Example

In a typical holographic recording configuration, a 50 Vμm^{-1} poling field is applied. Two coherent writing beams form an angle of 6° and their bisector is incident at 60° from normal incidence and the poling field direction. Refraction of the beams leads to a grating vector which is tilted $\phi = 58$° from the normal. The probe wave in a four wave mixing geometry is often counter-propagating to one of the writing beams and in this case has an electric field $\beta = 56$° from the poling field direction. From the theory of charge generation presented in chapter 4 the saturated space-charge field will not exceed the product of the projection of the poling field and the intensity contrast, corrected for field-dependent photogeneration,

$$E_{sc}\Big|_{s\propto E^p} \leq \frac{mE_0' \cos\phi}{1 + p\cos^2\phi} \,. \tag{6.46}$$

For contrast $m \to 1$ the first Fourier component of the modulated squared field is about 440 $(V\mu m^{-1})^2$. The electro-optic effect in the hologram will therefore be that equivalent to the application of around 21 $V\mu m^{-1}$ in an ellipsometric measurement of birefringence over a similar time scale. A material which exhibits a birefringence of 2.2×10^{-4} due to an applied field of 21 $V\mu m^{-1}$ will also have a holographic contrast of 2.2×10^{-4} for p-polarized diffraction in the above hologram recording experiment at a poling field of 50 $V\mu m^{-1}$.

A marked increase in the holographic contrast will be obtained if the orientations of the electric fields of the probe wave and the space-charge field are aligned better with the poling field.

Dye figure-of-merit for reorientational and Pockels effects

The C and A coefficients used to describe the electro-optic effect may be expressed to first order as

$$C = \frac{N}{45kT}\left(\frac{2\Delta\alpha\mu^2}{kT} + 9\beta\mu\right) \tag{6.47}$$

and

$$A = \frac{N}{45kT}\left(\frac{-\Delta\alpha\mu^2}{kT} + 3\beta\mu\right) \tag{6.48}$$

where the notation for polarizability anisotropy $\alpha_{//} - \alpha_{\perp} = \Delta\alpha$ has been adopted.

The C coefficient provides a simple figure-of-merit (FOM) that has been used to guide the development of improved chromophores for efficient photorefractive composites.[4] The figure identified for optimization is

$$F.O.M._C = \left(\frac{2\Delta\alpha\mu^2}{kT} + 9\beta\mu\right)M^{-1} \tag{6.49}$$

where the faster reorientation of a lighter chromophore has been accounted for by division by the molar mass, M.

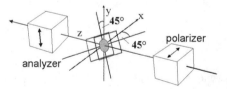

Figure 6.5 The ellipsometer is a device to measure the change in the state of polarization of light passed through a test device. The transmission through the ellipsometer is measured as the total transmitted intensity detected relative to the intensity incident on the sample.

The holographic contrast observed during a photorefraction experiment is proportional to $(C - A)$ and accordingly an alternative FOM would be to maximize this, equivalent to maximization of

$$F.O.M._{(C-A)} = \left(\frac{2\Delta\alpha\mu^2}{kT} + 4\beta\mu \right) M^{-1} \tag{6.50}$$

where the only difference is a reduced emphasis on the Pockels molecular hyperpolarizability, β.

Ellipsometry

The ellipsometer (Figure 6.5) is a reliable measure of the birefringence of a photorefractive composite when an electric field is applied to it. This provides the basic information needed to predict the holographic contrast in a photorefractive polymer.[5,6] The ellipsometer measures the field-induced difference in refractive index for light polarized perpendicular and parallel to the known applied field, $n_e - n_o$, where

$$n_e - n_o = \frac{1}{2n}(C - A)E^2 \tag{6.51}$$

and therefore the value of $C - A$ is measured directly. In order to distinguish between C and A, it is necessary to use microscopic parameters for the molecular dipoles. The ratio

$$\frac{A}{C - A} = \frac{3kT\beta - \mu\Delta\alpha}{6kT\beta + 3\mu\Delta\alpha} = \frac{\Delta n_o}{n_e - n_o} \tag{6.52}$$

is close to $-1/3$ for most materials which exhibit molecular reorientation during the photorefractive process. Materials with $T \ll T_g$ will be dominated by the Pockels effect and the ratio rises to $+1/2$ as the contribution due to $\Delta\alpha \to 0$.

References

1. Born, M. and Wolf, E., *Principles of Optics*, 3rd Ed., Oxford, Pergamon Press, 1965.
2. Sekkat, Z. and Knoll, W., Stationary state and dynamics of birefringence and nonlinear-optical properties induced by electric-field poling in polymeric films, *Ber. Bunsen. Phys. Chem.*, 98(10): 1231–1242, 1994.
3. Bant, S.P., Binks, D.J., and West, D.P., Full geometry dependence of index contrast in photorefractive polymer composites, *Appl. Opt.*, 41(11): 2111–2115, 2002.
4. Diaz-Garcia, M.A. *et al.*, *Chem. Mater.*, 11: 1784, 1999.
5. Shakos, J.D., Rahn, M.D., West, D.P., and Khand, K., Holographic index-contrast prediction in a photorefractive polymer composite based on electric-field-induced birefringence, *J. Opt. Soc. Am. B*, 17(3): 373–380, 2000.
6. Rahn, M.D., West, D.P., and Shakos, J.D., Photorefractive holographic contrast enhancement via increased birefringence in polymer composites containing electro-optic chromophores with different alkyl substituents, *J. Appl. Phys.*, 87(2): 627–631, 2000.

chapter seven

The dynamics of chromophore reorientation

The bulk of the index contrast in high-performance photorefractive polymers is due to the alignment of dipolar chromophores toward the local field direction within the material. The form of the index contrast transient observed is therefore strongly dependent on the dynamic reorientational response of the chromophores to changes in the local electric field. (The total field at any point in a photorefractive sample is the resultant of the field applied externally and the spatially varying space-charge field formed by the photorefractive effect.) A good understanding of chromophore rotational response not only allows a meaningful assessment of holographic formation speed and intercomparison of materials but it also allows the physical processes that underlie photorefraction, such as charge generation and transport, to be more readily investigated. Moreover, knowing the form of the index transient is useful in the rational design of many applications of photorefractive polymer composites. For instance, hologram multiplexing in a data storage device requires an understanding of the index growth dynamic so that hologram writing can be efficiently scheduled.

The rotational diffusion equation (RDE)

To a good approximation, the chromophores commonly used in low T_g photorefractive polymer composites can be considered to be elongated, cylindrically symmetrical dipoles that are free to rotate under the influence of an electric field, $\underline{E}$, and are subject to Brownian agitation. The orientation of an individual chromophore is represented by a unit vector, $\underline{n}$, that runs parallel to the long axis of the notionally cylindrical molecule; it is assumed that the dipole moment, μ , is also parallel to $\underline{n}$. The collective motion of a group of such dipoles can be described by the Smoluchowski rotational diffusion equation (RDE)[1]

$$\frac{\partial \Psi}{\partial t} = D\underline{R} \cdot \left[\underline{R}\Psi + \frac{\Psi}{kT}\underline{R}U \right]$$

(7.1)

where Ψ is the orientational distribution function, $U = -\mu \cdot E$ represents the interaction energy between the dipole and the electric field, the rotational operator is $\underline{R} = n \times \nabla$ and the diffusion coefficient, D, is related to the average rotation time, $\langle \tau_{rot} \rangle$ by[2]

$$D = \frac{1}{6\langle \tau_{rot} \rangle}$$

(7.2)

Clearly when considering rotational motion, it is convenient to use spherical coordinates (r, θ, ϕ), and in this case also to set the polar axis parallel to the local field direction. Hence, under spherically symmetric coordinates and assuming that Ψ varies only with polar angle, θ, and time, the RDE becomes

$$\frac{1}{D}\frac{\partial \Psi(\theta,t)}{\partial t} = \frac{1}{\sin\theta}\frac{\partial}{\partial\theta}\left(\sin\theta\left[\frac{\partial \Psi(\theta,t)}{\partial\theta} + \frac{1}{kT}\Psi(\theta,t)\frac{\partial U}{\partial\theta} \right] \right)$$

(7.3)

Legendre polynomials of the form $P_n(\cos\theta)$ often form a convenient basis set with which to represent the solutions to partial differential equations in spherical coordinates. Using these Ψ is now written as

$$\Psi(\theta) = \sum_{n=0}^{\infty} \frac{2n+1}{2} A_n P_n(\cos\theta)$$

(7.4)

The above equation acts as a more formal definition of the order parameters, A_n, that were introduced in the previous chapter. The nth-order parameter is the coefficient of the nth-order term in the Legendre series representation of the orientational distribution function. So, explicitly the order parameters can be written using the orthogonality of Legendre polymonials:

$$A_n = \int_{-1}^{+1} \Psi(\theta)P_n(\cos\theta).d(\cos\theta) = \langle P_n(\cos\theta) \rangle$$

(7.5)

Using this representation of $\Psi(\theta)$ the RDE is converted to a set of coupled, partial differential equations of the form

$$\frac{1}{D}\frac{\partial A_n}{\partial t} = -n(n+1)A_n + u\frac{n(n+1)}{2n+1}\left[A_{n-1}(t) - A_{n+1}(t)\right], \quad n = 0,1...\infty, \quad (7.6)$$

where $u = \mu E / kT$ is the balance parameter. Equations (7.6) are not analytically soluble in general but for typical values of $u (\sim 0.35$ for a 40 V/μm field) the values of A_n are negligible for $n > 2$; furthermore, since $P_0 = 1$ then $A_0 = 1$ also for a constant population of chromophores. Hence, by setting $A_n = 0$ for $n > 2$ a good representation is maintained while reducing the problem to a set of just two coupled first-order, inhomogeneous differential equations in the dependent variables A_1 and A_2:

$$\frac{1}{D}\frac{\partial A_1}{\partial t} = -2A_1 + u\frac{2}{3}\left[1 - A_2(t)\right] \tag{7.7}$$

$$\frac{1}{D}\frac{\partial A_2}{\partial t} = -6A_2 + u\frac{6}{5}A_1(t) \tag{7.8}$$

In (7.7) and (7.8) the presence of an electric field acts through the balance parameter to couple the two equations. The form of the solution to these equations depends strongly on the time-dependence of the diffusion coefficient, D.

Solution of the RDE in dispersion-free environments

For the purposes of this analysis, a chromophore is regarded to be rotating in a dispersion-free environment if the diffusion coefficient is a constant. In the dispersion-free case (7.7) and (7.8) have a solution that is bi-exponential in form for a field applied as a step-function:[3]

$$A_1(t) = A_1^\infty \left\{1 + \frac{\tau_1}{\tau_2 - \tau_1}\exp(-t / \tau_1) - \frac{\tau_2}{\tau_2 - \tau_1}\exp(-t / \tau_2)\right\}$$

$$-2D\frac{u}{3}\frac{\tau_2\tau_1}{\tau_2 - \tau_1}\left[\exp(-t / \tau_1) - \exp(-t / \tau_2)\right]$$

$$\tag{7.9}$$

$$A_2(t) = A_2^\infty \left\{1 + \frac{\tau_1}{\tau_2 - \tau_1}\exp(-t / \tau_1) - \frac{\tau_2}{\tau_2 - \tau_1}(-t / \tau_2)\right\}$$

where

$$\tau_{1,2} = \left[2D\left(2 \mp \sqrt{1 - u^2 / 5} \right) \right]^{-1} \tag{7.10}$$

and

$$A_1^\infty = \frac{5u}{15 + u^2}$$

$$\tag{7.11}$$

$$A_2^\infty = \frac{u^2}{15 + u^2}$$

The dynamic behavior of the second-order parameter transient is of particular relevance to the photorefractive effect because, as noted in the previous chapter, it is proportional to the field-induced birefringence, $n_e - n_o$, in the material, which can be measured directly by transmission ellipsometry. For low T_g composites, the induced birefringence is largely due to chromophore reorientation, which allows Pockel's effect to be neglected; further noting that typically $A_2^\infty \approx u^2/15$, the general relationship between the second-order parameter dynamic and the birefringence transient can be written simply as

$$n_e - n_o = \frac{N\Delta\alpha}{2n} A_2(t) \tag{7.12}$$

which for the present case gives

$$n_e - n_o = \frac{(C - A)}{2n} E_0^2 \left\{ 1 + \frac{\tau_1}{\tau_2 - \tau_1} \exp\left(-t / \tau_1\right) - \frac{\tau_2}{\tau_2 - \tau_1} \exp\left(-t / \tau_2\right) \right\} \tag{7.13}$$

The above equation corresponds to the birefringence transient induced by the application of an external field as step-function. Monitoring the resulting birefringence transient in a transmission ellipsometry experiment thus affords a method of verifying this model of chromophore reorientation in photorefractive polymer composites directly, deconvolved from the other processes involved in photorefraction. Example transients are given in Figure 7.1.

For relaxation the field is removed hence, $u = 0$, so the RDE becomes:

$$\frac{\partial A_1}{\partial t} = -2A_1 D \tag{7.14}$$

with

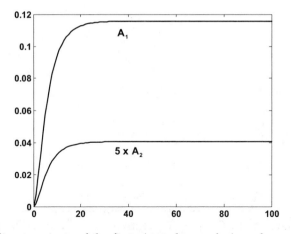

Figure 7.1 Poling transients of the first, A_1, and second, A_2, order parameters for a dispersion-free material. The transients were calculated for a diffusion constant of $D = 0.1$ (msec)$^{-1}$ and a balance parameter of $u = 0.35$, which are typical experimental values. For clarity, A_2 has been scaled up by a factor of 5.

$$\frac{\partial A_2}{\partial t} = -6A_2 D \tag{7.15}$$

which has the following solutions (again using (7.12) to express in terms of field-induced birefringence):

$$A_1(t) = A_1^0 e^{-2D_0 t}$$

$$n_e - n_o = \frac{N\Delta\alpha}{2n} A_2(t) = (n_e - n_o)_0 e^{-6D_0 t} \tag{7.16}$$

where A_1^0 and $(n_e - n_o)_0$ are the steady-state values of the first-order parameter and birefringence before the poling field is removed. This single exponential result is analogous to that first famously found by Peter DeBye[4,5] for dipoles relaxing in dilute liquids. Figure 7.2 shows some typical relaxation transients.

Hence, in the case of a dispersion-free environment, order parameter dynamics are exponential or bi-exponential in form and are characterized by definite lifetimes. This type of transient has been observed to be a good description of the poling and relaxation of dilute concentrations of dipoles in low viscosity host materials. However, it has long been recognized[6] that it does not represent well the transients observed at high concentrations or in amorphous media. Chromophores can constitute up to ~50 wt.% in some photorefractive polymer composites and these can have a glass transition temperature similar to room temperature. It is not surprising then that many such composites exhibit nonexponential order parameter transients.

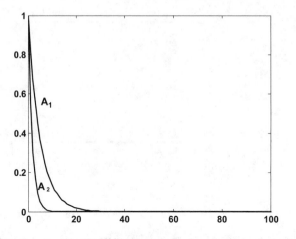

Figure 7.2 Relaxation transients of the first, A_1, and second, A_2, order parameters for a dispersion-free medium. The transients were calculated for a diffusion constant of $D = 0.1$ (msec)$^{-1}$. The initial value of both A_1 and A_2 was set to unity.

Solutions of RDE in dispersive environments

The average rotation time for a collection of dipoles, $\langle \tau_{rot} \rangle$ is given by

$$\langle \tau_{rot} \rangle = \int_0^\infty \varphi(\tau)\tau.d\tau \qquad (7.17)$$

where τ is the rotation time of an individual dipole and $\varphi(\tau)$ is the rotation time distribution function. If the above integral converges then $\langle \tau_{rot} \rangle$ and consequently the diffusion coefficient are constants. In this case the exponential form of transients described above may be observed and dipole rotation in the polymer environment may be considered dispersion-free. However, in disordered materials such as amorphous polymer composites each dipole is likely to be situated in a unique local molecular environment; some dipoles will be able to move relatively freely and will respond rapidly to changes in the local field while others will have their movement restricted and take much longer to reorient. It is possible that the resulting rotation time distribution function is such that the integral in (7.17) does *not* converge. Mathematically, this means that first moment about zero of $\varphi(\tau)$, and hence the mean rotation time, is infinite. Microscopically, a divergent first moment means that there will always be some dipoles whose extremely long rotation time outweighs their scarcity. Hence, any attempt to measure the mean rotation time will give a value that depends on the length of time over which the experiment was conducted; observing for a longer time will always result in a larger value of $\langle \tau_{rot} \rangle$. To incorporate this phenomenon (7.17) can be rewritten to include this dependence on the time over which the averaging has been done:

$$\langle \tau_{rot} \rangle = \int_0^t \varphi(\tau)\tau.d\tau = f(t) \qquad (7.18)$$

In this analysis, materials for which $\langle \tau_{rot} \rangle$ is a function of time are regarded as dispersive. It is the effective time-dependence of the diffusion coefficient, D, that causes nonexponential dynamic behavior typical of dispersive materials.

Nonexponential index contrast transients are observed commonly for dipolar chromophores in photorefractive composites under the influence of fields applied as step-functions. This can be taken as an indication that $\langle \tau_{rot} \rangle$ is a function of time. What then is the appropriate functional form for the time dependence?

Dispersive charge transport in amorphous materials has been described successfully[7,8] (see Chapter 5) by a hopping time distribution, $\psi(\tau_{hop})$ of the form

$$\psi(\tau_{hop}) \propto \frac{1}{\tau_{hop}^{1\mp\alpha}} \qquad (7.19)$$

where α is a parameter that describes the degree of disorder in the material; for complete order $\alpha = 1$ and for complete disorder $\alpha = 0$. (The transition from the $-\alpha$ to the $+\alpha$ form happens when a certain fraction of the moving charges reach the target electrode.) This simple type of distribution has the qualities necessary to be consistent with the observed behavior of dipole rotation in amorphous media. It produces a nonconverging mean rotation time and a nonexponential rotational transient, except in the case of complete order, $\alpha = 1$ where it results in a constant $\langle \tau_{rot} \rangle$ value and exponential dynamic behavior. Charge transport in amorphous materials is largely dependent on the distribution of waiting times between hops, i.e., $\psi(\tau_{hop})$. By analogy then, chromophore reorientation may be similarly dependent on the waiting time between rotations, $\varphi(\tau)$, making the reciprocal power law form for the diffusion process a reasonable one to try. There is no event equivalent to charge reaching an electrode and leaving the material so for dipole rotation there is not expected to be a transition from the $-\alpha$ to the $+\alpha$ form.

Complete disorder case

In the first instance complete disorder is assumed for simplicity, $\alpha = 0$. Under this assumption the mean rotation time $\langle \tau_{rot} \rangle \propto t$ and the time-dependent diffusion coefficient $D(t) = D_0/t$ for $t \geq 1$, where a diffusion constant, D_0, has been introduced. The condition $t \geq 1$ has been included to ensure that the resulting order parameter transients remain finite and is only necessary when $\alpha = 0$. With less than complete disorder, $\alpha > 0$, which is more physical, the transients are nonsingular even at $t = 0$. The condition is not a practical problem because the units of time used can be set to be arbitrarily small so that the time $t = 1$ can be as close to zero as required; the magnitude of D_0 depends on the units of time used (even though it is itself dimensionless).

For the rotation of dipoles in a dispersive medium of complete disorder the RDE becomes:

$$\frac{t}{D_0} \frac{\partial A_1}{\partial t} = -2A_1 + u \frac{2}{3}\left[1 - A_2(t)\right] \tag{7.20}$$

with

$$\frac{t}{D_0} \frac{\partial A_2}{\partial t} = -6A_2 + u \frac{6}{5} A_1(t) \tag{7.21}$$

This form of the RDE readily is solved for poling by a field applied as a step-function by making the substitution $t = e^z$. This renders (7.20) and (7.21) into the same functional form as (7.7) and (7.8), respectively, but in the independent variable z rather than t. Hence, the solution in terms of the variable z is identical to (7.9) and can be made explicit in terms of t by using the reverse substitution $z = \ln(t)$ to give a bi-power law dynamic[9,10]

$$A_1(t) = A_1^\infty \left\{ 1 + \frac{\tau_1}{\tau_2 - \tau_1} t^{s_1} - \frac{\tau_2}{\tau_2 - \tau_1} t^{s_2} \right\} - 2D_0 \frac{u}{3} \frac{\tau_2 \tau_1}{\tau_2 - \tau_1} \left[t^{s_1} - t^{s_2} \right]$$

$$n_e - n_o = \frac{N\Delta\alpha}{2n} A_2(t) \tag{7.22}$$

$$= \frac{N\Delta\alpha}{2n} A_2^\infty \left\{ 1 + \frac{\tau_1}{\tau_2 - \tau_1} t^{s_1} - \frac{\tau_2}{\tau_2 - \tau_1} t^{s_2} \right\}$$

where the two field-dependent time constants

$$\tau_{1,2} = \left[2D_0 \left(2 \mp \sqrt{1 - u^2/5} \right) \right]^{-1} = -s_{1,2}^{-1} \tag{7.23}$$

Examples of dispersive transients for the first-order parameters are shown in Figure 7.3.

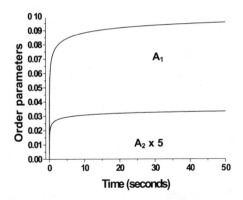

Figure 7.3 Poling transients of the first, A_1, and second, A_2, order parameters for a dispersive medium. The transients were calculated for a diffusion constant of $D_0 = 0.1$ and a balance parameter of $u = 0.35$, which are typical experimental values. For clarity, A_2 has been scaled up by a factor of 5. (After Binks, D.J., Khand, K., and West, D.P., *J. Appl. Phys.*, 89: 231, 2001.)

Unlike the equivalent dynamic for a dispersion-free medium, this type of transient does not possess a characteristic timescale. It is common practice to characterize a transient of unknown functional form empirically by a multi-exponential fit.[11] In the case of literature about photorefractive polymer composites this empirical characterization process has often utilized a bi-exponential[12,13] or stretched (Kohlrausch-Williams-Watts) exponential,[14] with the bi-exponential fit being most common. However, as illustrated in Figure 7.4, the bi-power law function given in (7.22) describes the observed transients better than does a bi-exponential (similar to (7.9)), resulting in a better fit and reduced residuals.

Similarly, the dispersive relaxation transients become:

$$A_1(t) = A_1^0 t^{-2D_0}$$

$$n_e - n_o = \frac{N\Delta\alpha}{2n} \qquad A_2(t) = (n_e - n_o)_0 \, t^{-6D_0}$$

(7.24)

where A_1^0 and $(n_e - n_o)_0$ are initial values of the first-order parameter and the birefringence (i.e., at $t = 1$). Figure 7.5 shows some example relaxation dynamics for dipoles in a dispersive medium.

Since $\tau_{1,2}$ are a very weak function of u for typical material parameters and experimental fields, the normalized form of the second-order parameter dynamic, $A_2(t)/A_2^\infty$, is substantially determined by a single material parameter: the diffusion constant, D_0.

Even though a characteristic rotation time cannot be defined due to the nonconvergence of the integral in (7.17), over a fixed measurement time materials with higher values of D_0 will attain a greater fraction of their final

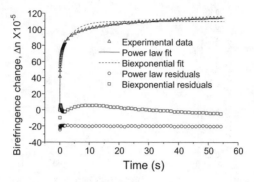

Figure 7.4 Transmission ellipsometer data showing the field-induced birefringence of a sample containing 40 wt.% EHDNPB in a host consisting of 45 wt.% PVK and 15 wt.% ECZ. The sample had a thickness of 59 μm and a 2.36 kV step function was applied across the electrodes; thus the field applied was 40 V.μm^{-1}. Both a power law and a bi-exponential fit are shown as well as the corresponding residuals. The power law fit resulted in a value of $D_0 = 0.0722 \pm 1$ and the bi-exponential fit gave $\tau_{fast} = 63 \pm 1$ ms. For clarity, only every 20th point has been plotted and the residuals for the power law fit have been offset by $\Delta n = -20 \times 10^{-5}$. (From Binks, D.J., Khand, K., and West, D.P., *J. Opt. Soc. Am. B*, 18: 308, 2001. With permission.)

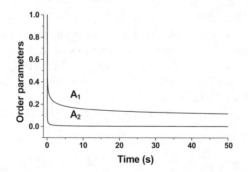

Figure 7.5 Relaxation transients of the first, A_1- and second- A_2, order parameters for a dispersive medium. The transients were calculated for a diffusion constant of $D_0 = 0.1$. The initial value of both A_1 and A_2 was set to unity. (From Binks, D.J., Khand, K., and West, D.P., *J. Appl. Phys.*, 89: 231, 2001. With permission.)

birefringence than low D_0 materials, as is illustrated in Figure 7.6. Hence, the diffusion constant can be regarded as a measure of the rotational freedom of the chromophores in the amorphous host.

The rotational freedom of the chromophores can be increased by altering the composition of the polymer; in particular, the addition of a plasticizing agent increases the value of D_0. In the case of PVK-based composites, the replacement of some of the PVK by its monomer, ethylcarbazole (ECZ), plasticizes the composite and enhances the diffusion constant, as shown in Figure 7.7. Plasticization can also be achieved in PVK-based composites by

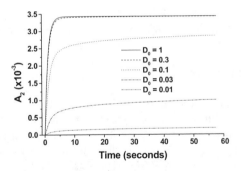

Figure 7.6 Second-order parameter transient for various values of diffusion constant, D_0 (millisecond timescale). The calculations utilized a space-charge field rise time of $\tau_{sc} = 1$ second and an effective applied field of 26 V/μm. (From Binks, D. J. and West, D. P., *J. Opt. Soc. Am. B*, 19: 2349, 2002. With permission.)

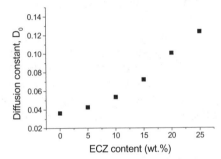

Figure 7.7 Variation of the diffusion constant, D_0, (millisecond timescale) with plasticizer content. Each material tested was identical except for the proportions of ECZ, which acted as the plasticizer, and the PVK that it replaced. The chromophore concentration was 40 wt.% and the applied field was 40 V.μm⁻¹ in each case. The errors calculated from the fit to the transmission ellipsometer data are smaller than the symbols used. (From Binks, D.J., Khand, K., and West, D.P., *J. Opt. Soc. Am. B*, 18: 308, 2001. With permission.)

increasing the fraction of chromophore used[13,15] for instance, increasing the fraction of dye from 40 wt.% to 47.5 wt.% resulted in a 50% enhancement in D_0 in one case.[15] However, it should be noted that too much plasticization causes phase separation in polymer composites, which effectively destroys the useful photorefractive properties of the material.

Arbitrary disorder

It is more physical that neither complete order nor complete disorder is assumed, $0 < \alpha < 1$, when the disorder parameter is added to the description of chromophore reorientation. In this case, the diffusion coefficient becomes $D(t) = D_0/t^{1-\alpha}$ but otherwise the analysis proceeds in largely the same manner as for the complete disorder, $\alpha = 0$, case. Equations (7.7) and (7.8) can again be solved by changing to a new variable, z, where in general

$$z = \int D(t)dt \qquad (7.25)$$

They are thus rendered in a form that has the same solutions as the dispersion-free case but in the new variable; the transient is found by changing the solutions to a form that is explicitly in terms of the time, t. In the current case of intermediate disorder this yields the following form for the chromophore response to a step-function applied field:

$$n_e - n_o = \frac{N\Delta\alpha}{2n} A_2^\infty \left\{ 1 + \frac{\tau_1}{\tau_2 - \tau_1} \exp\left(-\frac{t^\alpha}{\alpha\tau_1}\right) - \frac{\tau_2}{\tau_2 - \tau_1} \exp\left(-\frac{t^\alpha}{\alpha\tau_2}\right) \right\} \qquad (7.26)$$

An example fit of (7.26) to transmission ellipsometry data is shown in Figure 7.8.

Functions of the form $\exp(x^a)$ are known as stretched exponentials and are used to describe many properties of disordered materials.[2,16] This description is usually empirical although there have been a few stochastic models that predict how stretched exponential behavior might be caused by a power law distribution of waiting times.[17,18] It is interesting to note that a bi-stretched exponential function is derived above simply by assuming a power law form for φ.

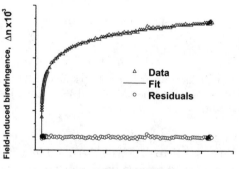

Figure 7.8 Result of a transmission ellipsometer experiment for a sample containing 40 wt.% of the chromophore EHDNPB in a PVK host medium. The sample had a thickness of 50 μm and a 2.0 kV step function was applied across the electrodes. A data point was taken every 2 ms for the first 0.8 s, to resolve the fast-changing initial part of the transient, and then every 60 ms for the later part. For clarity, only every fifth point is plotted. (From Binks, D.J. and West, D. P., *J. Chem. Phys.*, 115: 1060, 2001. With permission.)

It has been shown[19] that (7.26), which allows for partial disorder, fits observed birefringence transients better than (7.22), which assumes complete disorder. It is more plausible that disorder is neither absent nor complete. The intermediate disorder model is also more mathematically satisfactory at $t = 0$. However, one important advantage that the complete disorder model does have over the partial disorder description is that the form of the transient for complete disorder is amenable to analysis by Laplace transformation. As will be described in the next section, this allows it to be used in the calculation of an analytical form for the index contrast dynamic observed in photorefractive experiments.

Index contrast growth in photorefractive polymers

The field-induced birefringence transient observed in a photorefractive polymer composite during transmission ellipsometry is due mostly to chromophore reorientation. When the field is applied as a step function it is described well by (7.22). In many applications of photorefraction in polymeric composites, a steady applied poling field is used. During the buildup of the photorefractive effect the orientation of the chromophores is altered by the resultant of this steady applied external field and a space-charge field, E_{sc}, which grows exponentially according to the standard theory of photorefractivity[20,21] as

$$E_{sc}(t) = E_{sc,f}\left[1 - \exp\left(-\frac{t}{\tau_{sc}}\right)\right] \tag{7.27}$$

where $E_{sc,f}$ is the final value of the space-charge field and τ_{sc} is the space-charge field rise time. We use the concept of a Laplacian transfer function as the response of a system to an impulse. The index contrast transient observed during the photorefractive effect will be the convolution of the exponentially growing space-charge field and the transfer function for chromophore rotation.

Numerical calculation of index contrast growth

Numerical convolution yields some important insights into the most commonly used method of interpretation of the index contrast transients observed during photorefractive experiments.[15] The faster of two time constants from an empirical bi-exponential fit to the index-contrast transient has been the predominant method of assessing the rate of space-charge field formation for the photorefractive effect in polymers. However, fitting a bi-exponential function to the numerical convolution reveals that this approach can lead to misleading results. Figure 7.9 shows the relationship between the faster empirical time constant and the underlying space-charge

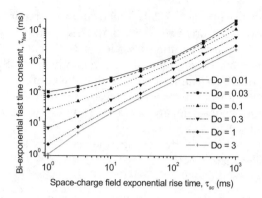

Figure 7.9 The relationship between space-charge field rise time and fast bi-expo-nential time constant for typical values of the diffusion constant, D_0 (millisecond timescale). An applied field value of 40 V.µm^{-1} was used in the calculations and the data was generated for a period of 20 seconds. (From Binks, D.J., Khand, K., and West, D.P., *J. Opt. Soc. Am. B*, 18: 308, 2001. With permission.)

field rise-time. Their values are often very different and the ratio between them depends on the diffusion constant of the material (but not on the applied field strength[15]). Hence the assessment of photorefractive response speed by the empirical fitting of a bi-exponential is not a reliable basis for the comparison of materials. In contrast, knowing the exact form of the index-contrast transient including the influence of the diffusion constant would allow response speeds to be assessed so that materials can be com-pared without any ambiguity. Such an exact form can be found by the analytical convolution of the rotational transfer function of the chro-mophores and an exponentially growing driving field.

Analytical form for index-contrast growth

The convolved rotational response of the chromophores to an arbitrary driv-ing field function can be found by multiplying the Laplace transforms of the *signal* (the driving field) and *transfer* functions. The transfer function can be found by dividing the transform of the response to a known signal by the transform of that signal. In this case, the response of the chromophores to a step-function signal is known (equation 7.22). By standard Laplace transform results,[22] the second-order parameter expressed in transform space is

$$A_2(p) = \frac{1}{15}\left(\frac{\mu}{kT}\right)^2 E_0^2 \left\{\frac{1}{p} + \left(\frac{\tau_1}{\tau_2 - \tau_1}\right)\frac{\Gamma(s_1 + 1)}{p^{s_1+1}} - \left(\frac{\tau_2}{\tau_2 - \tau_1}\right)\frac{\Gamma(s_2 + 1)}{p^{s_2+1}}\right\}$$

$$(7.28)$$

where p is the transform variable and Γ is the gamma function. Strictly, the parameters s_i where $i = 1,2$ are field-dependent. However, for typical materials they are only very weak functions of field and they have been regarded as constants in the above transformation,

$$\tau_{1,2} = \left[2D_0\left(2 \mp 1\right)\right]^{-1} = -s_{1,2}^{-1} \qquad (7.29)$$

Laplace transforms are only defined for times $t \geq 0$ so the transform of a constant field, E_0, and a field applied as a step function at $t = 0$ are identical, $E = E_0/p$. Hence we are free to view the signal function either as the square of a step function, $(E_0/p)^2$, or as the product of a step function and a constant, $E_0 \cdot (E_0/p)$, as long as we are consistent. Physically, there is not expected to be any significant difference between the two cases since typical poling fields only slightly perturb the angular distribution of chromophores (for example,[23] a field of 50 V/µm only changes the average angle between the field direction and chromophore axis by ~1°). We wish to emphasize the correspondence of the resulting transfer function with photorefractive index contrast growth, which depends on the product of a constant field and a growing space-charge field. We choose the latter viewpoint and hence divide (7.28) by signal function $E_0 \cdot (E_0/p)$ to get the transfer function

$$T(p) = A_2(p) \times \frac{p}{E_0^2}$$
$$= \frac{1}{15}\left(\frac{\mu}{kT}\right)^2 \left\{1 + \left(\frac{\tau_1}{\tau_2 - \tau_1}\right)\frac{\Gamma(s_1+1)}{p^{s_1}} - \left(\frac{\tau_2}{\tau_2 - \tau_1}\right)\frac{\Gamma(s_2+1)}{p^{s_2}}\right\} \qquad (7.30)$$

The transform, $S(p)$, of the product of a constant poling field and the exponentially growing space-charge (7.27) corresponds to the driving function in a photorefractive experiment and is given by standard results[22]

$$S(p) = E_0 E_{sc,f}\left[\frac{1}{p} - \frac{1}{p + \tau_{sc}^{-1}}\right] \qquad (7.31)$$

Thus, the transform of the second-order parameter transient, $A_2^{PR}(p)$, corresponding to chromophore reorientation in a photorefractive experiment is

$$A_2^{PR}(p) = S(p) \times T(p)$$

$$= \frac{E_{sc,f}E_0}{15}\left(\frac{\mu}{kT}\right)^2$$

$$\left\{\left[\left(\frac{1}{p} + \left(\frac{\tau_1}{\tau_2 - \tau_1}\right)\frac{\Gamma(s_1+1)}{p^{s_1+1}} - \left(\frac{\tau_2}{\tau_2 - \tau_1}\right)\frac{\Gamma(s_2+1)}{p^{s_2+1}}\right)\right.\right.$$

$$\left.\left.-\left(\frac{1}{p+\tau_{sc}^{-1}} + \left(\frac{\tau_1}{\tau_2 - \tau_1}\right)\frac{\Gamma(s_1+1)}{p^{s_1}\left(p+\tau_{sc}^{-1}\right)} - \left(\frac{\tau_2}{\tau_2 - \tau_1}\right)\frac{\Gamma(s_2+1)}{p^{s_2}\left(p+\tau_{sc}^{-1}\right)}\right)\right]\right\}$$

$$(7.32)$$

The analytical form of the transient itself is given by finding the inverse Laplace transform of (7.32); this can be done using standard results for all but the last two terms on the right-hand side. Appendix A details the explicit evaluation of the Bromwich integral necessary in finding the inverse transform of these final two terms. Hence, using the results of Appendix A and expressing the transient in terms of the experimentally observable holographic index contrast, $\Delta n_{1K}\big|_{p-pol}$ the relation[24]

$$\Delta n_{1K}\big|_{p-pol}(p) = k_{geom,PR}\frac{N\Delta\alpha}{2n}A_2^{PR}(t)$$

$$= k_{geom,PR}\frac{(C-A)E_{sc,f}E_0}{2n}$$

$$\left\{\left[\left(1+\left(\frac{\tau_1}{\tau_2 - \tau_1}\right)t^{s_1} - \left(\frac{\tau_2}{\tau_2 - \tau_1}\right)t^{s_2}\right)\right.\right.$$

$$\left.\left.-e^{-t/\tau_{sc}}\left(1+\frac{\gamma(s_1,-t/\tau_{sc})}{(\tau_1-\tau_2)}(-\tau_{sc})^{s_1} - \frac{\gamma(s_2,-t/\tau_{sc})}{(\tau_1-\tau_2)}(-\tau_{sc})^{s_2}\right)\right]\right\}$$

$$(7.33)$$

is found where $k_{geom,PR}$ encompasses all the geometrical factors in (6.45) and $\gamma(s_i, -t/\tau_{sc})$ with $i = 1,2$ is the incomplete gamma function, the definition* of which is

---

* Strictly, this definition is only valid for $s_i > 0$, but there is alternative definition in terms of a confluent hypergeometric function that is valid for all s_i. See I.S. Gradshteyn and I.M. Ryzhik, *Tables of Integrals, Series and Products*, 5th ed., A. Jeffrey ed. (Cambridge University, Cambridge), p. 949.

$$\gamma\left(s_i, -t/\tau_{sc}\right) = \int_0^{-t/\tau_{sc}} z^{s_i-1} \exp\left(-z\right) dz \qquad (7.34)$$

The numerical calculation of this quantity is discussed in Appendix B.
The form of the transient calculated by (7.33) for various typical values of
space-charge field rise time, τ_{sc}, applied field, E_0, and diffusion constant, D_0,
are shown in Figures 7.10 to 7.12. These figures show that the rise of the
index contrast transient is limited by either chromophore reorientation or
the rate of space-charge field formation, depending on the relative values of
D_0 and τ_{sc}. For instance, for $D_0 = 0.059$ (millisecond timescale) the index

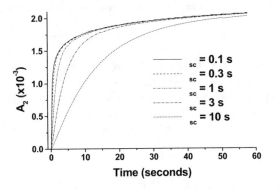

Figure 7.10 Second-order parameter transient for various values of space-charge field
rise time, τ_{sc}. The calculations utilized a diffusion constant of $D_0 = 0.059$ (millisecond
timescale) and an effective applied field of 26 V/μm. (From Binks, D.J. and West,
D.P., *J. Opt. Soc. Am. B*, 19: 2349, 2002. With permission.)

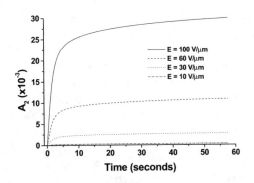

Figure 7.11 Second-order parameter transient for various values of applied field, E.
The calculations utilized a space-charge field rise time of $\tau_{sc} = 1$ second and a diffusion
constant of $D_0 = 0.059$ (millisecond timescale). (From Binks, D.J. and West, D.P., *J.
Opt. Soc. Am. B*, 19: 2349, 2002. With permission.)

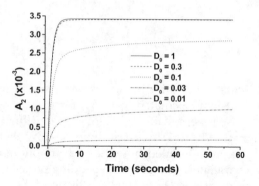

Figure 7.12 Second-order parameter transient for various values of diffusion constant, D_0 (millisecond timescale). The calculations utilized a space-charge field rise time of $\tau_{sc} = 1$ second and an effective applied field of 26 V/μm. (From Binks, D.J. and West, D.P., *J. Opt. Soc. Am. B*, 19: 2349, 2002. With permission.)

contrast response is limited by chromophore rotation for values of the τ_{sc} less than a few hundred milliseconds. If the rate of space-charge field formation becomes any slower then it starts to become the rate-limiting factor in the photorefractive response. In contrast, as shown by Figure 7.12, the applied electric field affects only the magnitude of the index contrast response and not its speed.

Uses for dynamic models of chromophore reorientation

In this chapter, the rotation of dipolar chromophores in a dispersive polymer environment has been used to describe and understand the refractive index dynamics measured in both transmission ellipsometry and photorefractive experiments. This has proved to be a very successful description of data observed across a wide range of experimental conditions both in ellipsometry and the four-wave-mixing (FWM) experiments that are used typically to study the growth of photorefractive index contrast. Moreover, the value of the diffusion constant, D_0, and the electro-optic coefficients, C and A, [(6.20), (6.21)] extracted from dynamic and steady-state transmission ellipsometry, respectively, can be used in the analysis of FWM leaving only the space-charge field rise time as the unknown parameter. This minimization of the number of fitting parameters increases the reliability of τ_{sc} extracted from FWM transients to reveal useful information about the photorefractive process. For instance, values of τ_{sc} extracted in this way were shown to be linearly dependent on the incident light intensity, as predicted by the standard theory of photorefractivity.[20,21] In contrast, previous studies[25] compared the bi-exponential fast time constant to light intensity and noted an apparent sublinear relationship, in apparent conflict with the standard model for a single trap level system and leading to speculation about shallow trap distributions in the materials. Measuring the variation of rise time with field is also useful because it has been shown[26] that photogeneration efficiency is the only field-dependent factor that contributes to the rate of space-charge field formation. Hence, this measurement provides a means to compare the effectiveness of different photorefractive sensitizing species.

References

1. Doi, M. and Edwards, S.F., Ch. 3, Brownian Motion, *The Theory of Polymer Dynamics,* Clarendon Press, Oxford, 1986.
2. Ngai, K.L., *J. Non-Cryst. Sol.* 275: 7, 2000.
3. Sekkat Z. and Knoll, W.,*Ber. Bunsenges. Phys. Chem.,* 98: 1231, 1994.
4. Debye, P., *Z. Phys,* 13: 97, 1913.
5. Debye, P., *Polar Molecules,* Lancaster, PA, 1929.
6. Williams, D.J., in Ch. II-7, Nonlinear Optical Properties of Guest-Host Polymer Structures, *Nonlinear Optical Properties of Organic Molecules and Crystals,* Chemla, D.S. and Zyss, J., Eds., Academic Press, New York, 1987.
7. Scher, H. and Montroll, E.W., *Phys. Rev. B,* 12: 2455, 1975.
8. Scher, H., Schlesinger, M.F., and Bendler, J.T., *Phys. Today,* 44: 26, 1991.
9. Binks, D.J. and West, D.P., *Appl. Phys. Lett.,* **77**: 1108, 2000.
10. Binks, D.J., Khand, K., and West, D.P., *J. Appl. Phys.,* 89: 231, 2001.
11. Hoffman, U., Schreiber, A., Haarer, D., Zilker, S.J., Bacher, A., Bradley, D.C.C., Redecker, M., Inbasekaran, M., Wu, W.W., and Woo, E.P., *Chem. Phys. Lett.,* 311: 41, 1999.
12. Bäuml, G., Schloter, S., Hofmann, U., and Haarer, D., *Opt. Commun.,* 154: 75, 1998.
13. Bittner, R., Bräuchle, C., and Meerholz, K., *Appl. Opt.,* 37: 2843, 1998.
14. Wright, D., Diaz-Garcia, M.A., Casperson, D.J., DeClue, M., and Moerner, W.E., *Appl. Phys. Lett.,* 73: 1490, 1999.
15. Binks, D.J., Khand, K., and West, D.P., *J. Opt. Soc. Am. B,* 18: 308, 2001.
16. Angell, C.A., Ngai, K.L., McKenna, G.B., McMillan, P.F., and Martin, S.W., *J. Appl. Phys.,* 88: 3113, 2000.
17. Klafter, J. and Schlesinger, M.F., *Proc. Natl. Acad. Sci. U. S. A.,* 83: 848, 1986.
18. Weron, K., *J. Phys. Condens. Matter,* 3: 9151, 1991.
19. Binks, D.J. and West, D. P., *J. Chem. Phys.,* 115, 1060, 2001.
20. Kukharerev, N.V., Markov, V.B., Soskin, M., and Vinetskii, V.L., *Ferroelectrics,* 22: 949, 1979.
21. Kukharerev, N.V., Markov, V.B., Soskin, M., and Vinetskii, V.L., *Ferroelectrics,* 22: 961, 1979.
22. Spiegel, M.R., *Theory and Problems of Laplace Transforms,* McGraw Hill, New York, 1965.
23. Binks, D.J. and West, D.P., *Appl. Phys. B,* 74: 279, 2002.
24. Binks, D.J. and West, D.P., *J. Opt. Soc. Am. B,* 19: 2349, 2002.
25. Moerner, W.E. and Silence, S.M., *Chem. Rev.,* 94: 127, 1994.
28. Binks, D.J. and West, D.P., *J. Chem. Phys.,* 115: 6760, 2001.

Appendix A

Inverse Laplace transform of $p^m/(p+g)$

The inverse Laplace transforms of the final two terms of Equation 7.34 are of the general form

Equation A.1

$$L^{-1}\left(\frac{p^m}{p+g}\right) = \frac{1}{i2\pi}\int_{c-i\infty}^{c+i\infty}\frac{p^m}{p+g}\exp(pt)dp = \frac{1}{i2\pi}\int_{c-i\infty}^{c+i\infty}f(p)dp$$

where $g(= 1/\tau_{sc})$ is a complex constant and $0 < m < 1$. Note the timescale of the experiment can always be chosen such that $m\ (= -s_i,\ i = 1,2)$, satisfies this condition.

Equation A.1 may be evaluated by considering the corresponding contour integral, which is equated to the right-hand side by the residue theorem:

Equation A.2

$$\frac{1}{i2\pi}\oint f(p)dp = \sum(residues)$$

Figure A.1 shows the complex plane including the integration path (ABC-DEF), a branch cut and the single pole at $p = -g$.
In polar coordinates (r, θ) the integrand is

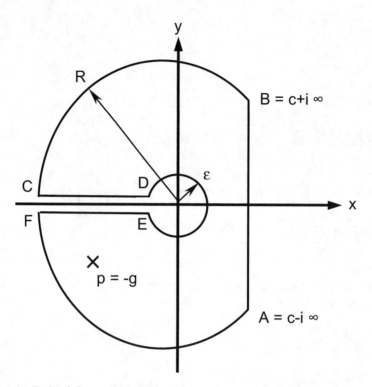

Figure A1 Path of the contour integration in the complex plane. The (×) represents the location of the pole at p = –g.

Equation A.3

$$f(r,\theta) = \frac{r^m \exp(im\theta)}{r \exp(i\theta) + g} \exp(r \exp(i\theta)t)$$

which tends to zero as $r \to 0$, and thus so does the integral. The integral also tends to zero in the limit of $r \to \infty$ because it satisfies the condition*

Equation A.4

$$\left| \frac{r^m \exp(im\theta)}{r \exp(i\theta) + g} \right| < \frac{M}{r^k} \qquad \text{as } r \to \infty$$

where M and k are constant greater than zero. Thus as $\varepsilon \to 0$ and $R \to \infty$, as shown in Figure A.1, then the integrals along *BC*, *DE* and *FA* tend to zero.

* Spiegel, M.R., *Theory and Problems of Laplace Transforms*, McGraw-Hill, New York, 1968, Chap. 7, p. 202.

The integrals on the paths *CD* and *EF* are both along the x-axis, either side of the branch cut, and hence their sum may be written in the limit of $\varepsilon \to 0$ and $R \to \infty$ by making the substitutions $p = x\exp(i\pi)$ and $p = x\exp(-i\pi)$, respectively, to give:

Equation A.5

$$\int_{EF} f(p)dp + \int_{CD} f(p)dp = \left(\exp(-im\pi) - \exp(im\pi)\right)\int_0^\infty \frac{x^m}{(x-g)}\exp(-xt)dx$$

$$= -2i\sin(m\pi)\int_0^\infty \frac{x^m}{(x-g)}\exp(-xt)dx$$

The solution to the integral on the right-hand side of Equation A.5 can be found in the literature*

Equation A.6

$$\int_0^\infty \frac{x^m}{(x-g)}\exp(-xt)dx = (-g)^m \exp(-gt)\Gamma(m+1)\Gamma(-m,-gt)$$

where $\Gamma(a)$ and $\Gamma(a,x)$ are the gamma function and the incomplete gamma function respectively, where

Equation A.7

$$\Gamma(a,x) = \int_x^\infty x^{a-1}\exp(-x)dx$$

$\Gamma(a)$ and $\Gamma(a,x)$ are simply related to $\gamma(a,x)$, as used in Chapter 7, by

Equation A.8

$$\Gamma(a) = \Gamma(a,x) + \gamma(a,x)$$

Hence, the inverse transform can now be written

* Gradshteyn, I.S. and Ryzhik, I.M., *Tables of Integrals, Series and Products*, 5th ed., Jeffrey, A., Ed., Cambridge University, Cambridge, p. 366.

Equation A.9

$$L^{-1}\left(\frac{p^m}{p+a}\right) = \frac{1}{i2\pi}\int_{AB} f(p)\,dp$$

$$= \sum(residue) - \frac{1}{i2\pi}\left[\int_{CD} f(p)\,dp + \int_{EF} f(p)\,dp\right]$$

$$= \sum(residue) + \frac{\sin(m\pi)}{\pi}(-g)^m \exp(-gt)\Gamma(m+1)\Gamma(-m,-gt)$$

There is only one pole within the integration path, at $p = -g$, and the residue due to such a simple pole is found by evaluating $f(p)(p+g)$ at $p = -g$. Thus, the required inverse transform can now be written, in terms of s_i and τ_{sc}, as

Equation A.10

$$L^{-1}\left(\frac{1}{p^{s_i}\left(p+\tau_{sc}^{-1}\right)}\right) = (-\tau_{sc})^{s_i}\exp\left(-\frac{t}{\tau_{sc}}\right)\left[1 - \frac{\sin(s_i\pi)}{\pi}\Gamma(1-s_i)\Gamma\left(s_i, -\frac{t}{\tau_{sc}}\right)\right]$$

If this expression is now multiplied by the coefficients that appear in Equation 7.32, and if Equation A.8 and the following properties of gamma functions* are applied

Equation A.11

$$\Gamma(s_i) = s_i\Gamma(s_i)$$

$$\Gamma(s_i)\Gamma(1-s_i) = \frac{\pi}{\sin(s_i\pi)}$$

then the expression becomes

Equation A.12

---

* Spiegel, M.R., *Mathematical Handbook of Formulas and Tables*, McGraw-Hill, New York, 1968, Chap. 16, p. 101.

$$\left(\frac{\tau_i}{\tau_2-\tau_1}\right)\Gamma\left(s_i+1\right)L^{-1}\left(\frac{1}{p^{s_i}\left(p+\tau_{sc}^{-1}\right)}\right)=\frac{\left(-\tau_{sc}\right)^{s_i}}{\left(\tau_1-\tau_2\right)}\exp\left(-\frac{t}{\tau_{sc}}\right)\gamma\left(s_i,-\frac{t}{\tau_{sc}}\right)$$

Appendix B

Numerical calculation of the incomplete gamma function

In order to evaluate the expression for photorefractive index contrast growth given in Equation 7.33, the value of the incomplete gamma function, $\gamma(s_i, -t/\tau_{sc})$, must be computed.

The incomplete gamma function may be calculated for any values of s_i and $(-t/\tau_{sc})$ by the following series*, truncated to obtain the required degree of accuracy:

Equation B.1

$$\gamma\left(s_i, -\frac{t}{\tau_{sc}}\right) = \Gamma(s_i)\left(-\frac{t}{\tau_{sc}}\right)^{s_i} \exp\left(\frac{t}{\tau_{sc}}\right) \sum_{n=0}^{\infty} \left(-\frac{t}{\tau_{sc}}\right)^n \frac{1}{\Gamma(s_i + n + 1)}$$

$$|t/\tau_{sc}| < \infty$$

Alternatively, for large values of $|t/\tau_{sc}|$ the asymptotic representation** can be used

---

* Abramowitz, M. and Stegun, I. *Handbook of Mathematical Functions*, 10th ed., National Institute of Standards and Technology, Washington, D. C., 1972, p. 262.
** Gradshteyn, I.S. and Ryzhik, I.M. *Tables of Integrals, Series and Products*, 5th ed., Jeffrey, A., Ed., Cambridge University, Cambridge, p. 951.

Equation B.2

$$\gamma\left(s_i,-\frac{t}{\tau_{sc}}\right)\approx\Gamma\left(s_i\right)-\left(-\frac{t}{\tau_{sc}}\right)^{s_i-1}\exp\left(\frac{t}{\tau_{sc}}\right)\sum_{n=0}^{N-1}(-1)^n\frac{\Gamma\left(1-s_i+n\right)}{\Gamma\left(1-s_i\right)}\left(-\frac{t}{\tau_{sc}}\right)^{-n}$$

$$\left|t/\tau_{sc}\right|\to\infty,\ -3\pi/2<\arg\left(-t/\tau_{sc}\right)<3\pi/2,\ N=1,2,\ldots$$

where the error will be of order $\left|t/\tau_{sc}\right|^{-N}$.

Note that the incomplete gamma function appears as a product with $\exp(-t/\tau_{sc})$ in Equation 7.33, which cancels out the exponentially growing term in both Equations B.1 and B.2.

Index

A

Amorphous composites, plasticized, 15
Amorphous organic materials
    hole drift velocity, 19
    photorefraction in, 9
Amorphous photorefractive materials
    cylindrical symmetry, 83
    dipolar molecules, 84
    hole transport, 10
Anion density changes, 23
Anion optical pattern, 12
Anisotropic dye, 35

B

Birefringence, 29, 34
    field-induced, 101
    induced, 100
    transients, 109
Bragg diffraction, 89
Bragg grating, 5
Braun model, 41, 43, 47
Bromwich integral, 112

C

Charge dissociation and electric field effects, 50
Charge generation, 17
    limit, 61
    limit model, 56–57, 58
    in organics, 41
Charge mobility and time-of-flight (TOF) measurements, 61
Charge movement in polymers, 5
Charge photogeneration, see Photogeneration

Chromophore reorientation
    dynamic models of, 110–111
    dynamics, 97ff
Coherence gate, 2
Complex rise-time, 53–54
Conduction current, 18
CT_1 state generation, 42
    interface effect on formation rate, 49
Cylindrical symmetry in amorphous media, 83

D

Dielectric tensor (ε), 82
Diffraction efficiency, 36
Diffusion current, 19
Diffusion field parameter, 24
Dipolar molecules, 84
Dispersion in transport, 66–67
Dispersive transport model of Scher and Montroll, 64–66
Dissociation of charges in organic materials, 50
Donor population, 21
Dopants, 56
DOS (density of states), 67–69
Drift mobility dependence, 77
Drift velocity, 61
Dye
    anisotropic, 35
    figure-of-merit, 94–95
    molecules
        asymmetric polarizability, 14
        shear viscosity, 14

E

Electrical neutrality, 20